Contents

46th Polish Mathematical Olympiad, 1994 ○ XLV Lithuanian Mathematical Olympiad, 1996
Vietnamese Mathematical Olympiad, 1996 ❑ 19th Austrian-Polish Mathematics Competition
37th IMO Turkish Team Selection Test ⋏ Australian Mathematical Olympiad, 1996
XXXIX Republic Competition of Mathematics in Macedonia ❑ Croatian Team Selection Test,1995
47th Polish Mathematical Olympiad, 1995 ○ Georg Mohr Konkurrencen I Matematik, 1996
St. Petersburg City Mathematical Olympiad, 1996 ❑ Ukrainian Mathematical Olympiad, 1996
Republic of Moldova XL Mathematical Olympiad,1996 ⋏ XII Italian Mathematical Olympiad,1996
31st Canadian Mathematical Olympiad, 1999 ○ South African Mathematical Olympiad, 1995
Taiwan Mathematical Olympiad, 1996 ❑ Croatian Mathematical Olympiad, 1995
13th Iranian Mathematical Olympiad, 1995 ⋏ XXXIII Spanish Mathematical Olympiad, 1996
Estonian Mathematical Contest, 1995 ○ Bi-National Israel-Hungary Competition, 1996
Finnish High School Mathematics Contest, 1997 ❑ Georgian Mathematical Olympiad, 1997
6th Taiwan Mathematical Olympiad, 1997 ⋏ 11th Iberoamerican Mathematical Olympiad, 1996
Chinese Mathematical Olympiad, 1997 ○ 48th Polish Mathematical Competition, 1997
18th Brazilian Mathematical Olympiad, 1996 ❑ Latvian Mathematical Olympiad, 1997 (1st TST)
Mathematical Olympiad in Bosnia and Herzegovina, 1997 ⋏ 5th Japan Mathematical Olympiad
23rd All Russian Olympiad, 1997 ❑ 20th Austrian-Polish Mathematical Competition, 1997
Israel Mathematical Olympiad, 1997 ○ Estonian Mathematical Olympiad, 1997
Ukrainian Mathematical Olympiad, 1997 ❑ 10th Irish Mathematical Olympiad, 1997
Hungary-Israel Bi-National Mathematical Competition, 1997 ⋏ Vietnamese Team Selection Test
7th Japan Mathematical Olympiad, 1997 ○ Vietnamese Mathematical Olympiad, 1997
38 IMO Turkey Team Selection Test, 1997 ❑ Chilean Mathematical Olympiad, 1994
24-th Spanish Olympiad ⋏ Bulgarian Mathematical Olympiad 1984
IMO Proposal by Poland (1985) ○ Annual Greek High School Competition 1983
IMO Proposal by Canada (1985) ❑ 19th Austrian Mathematical Olympiad, Final Round
United States Olympiad Student Proposals (1982) ⋏ Bulgarian Winter Competition 1983
British Mathematical Olympiad 1983 ❑ Asian Pacific Mathematical Olympiad 1989
Austrian-Polish Mathematics Competition 1982 ○ Leningrad High School Olympiad 1982
3rd Ibero-American Olympiad ❑ Spanish Mathematical Olympiad, First Round 1985
IMO Proposal by Ireland (1989) ⋏ Canadian Mathematics Olympiad 1991
Asian Pacific Mathematical Olympiad 1989 ○ Singapore Mathematical Competition 1988
Austrian-Polish Mathematical Competition 1986 ❑ USAMO Training Session 1986
Tenth Atlantic Provinces Mathematics Competition ⋏ 29th I.M.O. in Australia
XIV ALL UNION Mathematical Olympiad U.S.S.R. ○ 34th IMO Proposal by USA (1993)
43rd Mathematical Olympiad (1991-92) in Poland ❑ Nordic Mathematical Contest, 1992
10th Iranian Mathematical Olympiad ⋏ Canadian Mathematical Olympiad 1996
10th Iranian Mathematical Olympiad ❑ Turkish Mathematical Olympiad, 1993 (Selection Test)........

algebra
problems and solutions
from
Mathematical Olympiads

46th Polish Mathematical Olympiad, 1994

1

Find the number of those subsets of $\{1, 2, \ldots, 2n\}$ in which the equation $x + y = 2n + 1$ has no solutions.

solution page 57

XLV Lithuanian Mathematical Olympiad, 1996

1

Solve the following equation in positive integers:

$$x^3 - y^3 = xy + 61.$$

solution page 58

XLV Lithuanian Mathematical Olympiad, 1996

2

Sequences $a_1, \ldots, a_n, \ldots$ and $b_1, \ldots, b_n, \ldots$ are such that $a_1 > 0$, $b_1 > 0$, and

$$a_{n+1} = a_n + \frac{1}{b_n}, \quad b_{n+1} = b_n + \frac{1}{a_n}, \quad n \in \mathbb{N}.$$

Prove that

$$a_{25} + b_{25} > 10\sqrt{2}.$$

solution page 59

XLV Lithuanian Mathematical Olympiad, 1996

3

Two pupils are playing the following game. In the system

$$\begin{cases} *x + *y + *z = 0, \\ *x + *y + *z = 0, \\ *x + *y + *z = 0, \end{cases}$$

they alternately replace the asterisks by any numbers. The first player wins if the final system has a non-zero solution. Can the first player always win?

solution page 61

Vietnamese Mathematical Olympiad, 1996 (Category A)

1

Solve the system of equations:

$$\begin{cases} \sqrt{3x}\left(1 + \frac{1}{x+y}\right) = 2, \\ \sqrt{7y}\left(1 - \frac{1}{x+y}\right) = 4\sqrt{2}. \end{cases}$$

solution page 62

Vietnamese Mathematical Olympiad, 1996 (Category A)

6

We are given four non-negative real numbers a, b, c, d satisfying the condition:

$$2(ab + ac + ad + bc + bd + cd) + abc + abd + acd + bcd = 16.$$

Prove that:

$$a + b + c + d \geq \frac{2}{3}(ab + ac + ad + bc + bd + cd).$$

When does equality occur?

solution page 65

Vietnamese Mathematical Olympiad, 1996 (Category B)

6

Let x, y, z be three non-negative real numbers satisfying the condition: $xy + yz + zx + xyz = 4$.

Prove that: $x + y + z \geq xy + yz + zx$. When does equality occur?

solution page 66

19th Austrian-Polish Mathematics Competition, 1996

4

The real numbers x, y, z, t satisfy the equalities $x + y + z + t = 0$ and $x^2 + y^2 + z^2 + t^2 = 1$. Prove that $-1 \leq xy + yz + zt + tx \leq 0$.

solution page 67

19th Austrian-Polish Mathematics Competition, 1996

6

Natural numbers k, n are given such that $1 < k < n$. Solve the system of n equations

$$x_i^3 \cdot (x_i^2 + x_{i+1}^2 + \cdots + x_{i+k-1}^2) = x_{i-1}^2 \quad \text{for} \quad 1 \leq i \leq n,$$

with n real unknowns x_1, x_2, ..., x_n. Note: $x_0 = x_n$, $x_{n+1} = x_1$, $x_{n+2} = x_2$, and so on.

solution page 68

19th Austrian-Polish Mathematics Competition, 1996

7

Show that there do not exist non-negative integers k and m such that $k! + 48 = 48(k + 1)^m$.

solution page 69

37th IMO Turkish Team Selection Test

3

Given real numbers $0 = x_1 < x_2 < \cdots < x_{2n} < x_{2n+1} = 1$ with $x_{i+1} - x_i \leq h$ for $1 \leq i \leq 2n$, show that

$$\frac{1 - h}{2} < \sum_{i=1}^{n} x_{2i}(x_{2i+1} - x_{2i-1}) \leq \frac{1 + h}{2}.$$

solution page 70

37th IMO Turkish Team Selection Test

6

For which ordered pairs of positive real numbers (a, b) is the limit of every sequence (x_n) satisfying the condition

$$\lim_{n \to \infty} (a x_{n+1} - b x_n) = 0 \qquad (1)$$

zero?

solution page 71

Australian Mathematical Olympiad, 1996

5

Let $a_1, a_2, \ldots, a_n$ be real numbers and s, a non-negative real number, such that

(i) $a_1 \leq a_2 \leq \cdots \leq a_n$;   (ii) $a_1 + a_2 + \cdots + a_n = 0$;

(iii) $|a_1| + |a_2| + \cdots + |a_n| = s$.

Prove that

$$a_n - a_1 \geq \frac{2s}{n}.$$

solution page 72

47th Polish Mathematical Olympiad, 1995

3

Let $n \geq 2$ be a fixed natural number and let $a_1, a_2, \ldots, a_n$ be positive numbers whose sum equals 1.

(a) Prove the inequality

$$2 \sum_{i<j} x_i x_j \leq \frac{n-2}{n-1} + \sum_{i=1}^{n} \frac{a_i x_i^2}{1 - a_i}$$

for any positive numbers $x_1, x_2, \ldots, x_n$ summing to 1.

(b) Determine all n-tuples of positive numbers $x_1, x_2, \ldots, x_n$ summing to 1 for which equality holds.

solution page 73

XXXIX Republic Competition of Mathematics in Macedonia (Class I)

1

The sum of three integers a, b and c is 0. Prove that $2a^4 + 2b^4 + 2c^4$ is the square of an integer.

solution page 74

XXXIX Republic Competition of Mathematics in Macedonia (Class I)

2

Prove that if

$$a_0^{a_1} = a_1^{a_2} = \cdots = a_{1995}^{a_{1996}} = a_{1996}^{a_0}, \quad a_1 \in \mathbb{R}^*,$$

then

$$a_0 = a_1 = \cdots = a_{1996}.$$

solution page 75

XXXIX Republic Competition of Mathematics in Macedonia (Class II)

1

Prove that for positive real numbers a and b

$$2 \cdot \sqrt{a} + 3 \cdot \sqrt[3]{b} \geq 5 \cdot \sqrt[5]{ab}.$$

solution page 76

XXXIX Republic Competition of Mathematics in Macedonia (Class II)

4

Find the biggest value of the difference $x - y$ if $2 \cdot (x^2 + y^2) = x + y$.

solution page 77

XXXIX Republic Competition of Mathematics in Macedonia (Class III)

1

Solve the equation $x^{1996} - 1996x^{1995} + \cdots + 1 = 0$ (the coefficients in front of $x, \ldots, x^{1994}$ are unknown), if it is known that its roots are positive real numbers.

solution page 78

Georg Mohr Konkurrencen I Matematik, 1996

2

Determine all triples (x, y, z), satisfying

$$\begin{aligned} xy &= z, & (1) \\ xz &= y, & (2) \\ yz &= x. & (3) \end{aligned}$$

solution page 79

St. Petersburg City Mathematical Olympiad, 1996 (Third Round)

1

Serge was solving the equation $f(19x - 96/x) = 0$ and found 11 different solutions. Prove that if he tried hard he would be able to find at least one more solution.

solution page 80

St. Petersburg City Mathematical Olympiad, 1996 (11th Grade)

1

It is known about real numbers $a_1, \ldots, a_{n+1}; b_1, \ldots, b_n$ that $0 \leq b_k \leq 1$ $(k = 1, \ldots, n)$ and $a_1 \geq a_2 \geq \cdots \geq a_{n+1} = 0$. Prove the inequality:

$$\sum_{i=1}^{n} a_k b_k \leq \sum_{k=1}^{[\sum_{j=1}^{n} b_j]+1} a_k.$$

solution page 81

Republic of Moldova XL Mathematical Olympiad, 1996 (10)

1

Let $n = 2^{13} \cdot 3^{11} \cdot 5^7$. Find the number of divisors of n^2 which are less than n and are not divisors of n.

solution page 85

Republic of Moldova XL Mathematical Olympiad, 1996 (10)

2

Distinct square trinomials $f(x)$ and $g(x)$ have leading coefficient equal to one. It is known that $f(-12) + f(2000) + f(4000) = g(-12) + g(2000) + g(4000)$. Find all the real values of x which satisfy the equation $f(x) = g(x)$.

solution page 84

Republic of Moldova XL Mathematical Olympiad, 1996 (11-12)

1

Prove the equality

$$\frac{1}{666} + \frac{1}{667} + \cdots + \frac{1}{1996} = 1 + \frac{2}{2 \cdot 3 \cdot 4} + \frac{2}{5 \cdot 6 \cdot 7} + \cdots + \frac{2}{1994 \cdot 1995 \cdot 1996}.$$

solution page 85

Republic of Moldova XL Mathematical Olympiad, 1996 (11-12)

2

Prove that the product of the roots of the equation

$$\sqrt{1996} \cdot x^{\log_{1996} x} = x^6$$

is an integer number and find the last four digits of this number.

solution page 86

Republic of Moldova XL Mathematical Olympiad, 1996 (10)

5

Prove that for all natural numbers $m \geq 2$ and $n \geq 2$ the smallest among the numbers $\sqrt[n]{m}$ and $\sqrt[m]{n}$ does not exceed the number $\sqrt[3]{3}$.

solution page 87

Republic of Moldova XL Mathematical Olympiad, 1996 (10)

6

Prove the inequality $2^{a_1} + 2^{a_2} + \cdots + 2^{a_{1996}} \leq 1995 + 2^{a_1 + a_2 + \cdots + a_{1996}}$ for any real non-positive numbers $a_1, a_2, \ldots, a_{1996}$.

solution page 88

Republic of Moldova XL Mathematical Olympiad, 1996 (11-12)

6

Solve in real numbers the equation

$$2x^2 - 3x = 1 + 2x\sqrt{x^2 - 3x}.$$

solution page 89

31st Canadian Mathematical Olympiad, 1999

5

Let x, y and z be non-negative real numbers satisfying $x + y + z = 1$. Show that

$$x^2 y + y^2 z + z^2 x \leq \frac{4}{27},$$

and find when equality occurs.

solution page 90

6

Ukrainian Mathematical Olympiad, 1996

The sequence $\{a_n\}$, $n \geq 0$, is such that $a_0 = 1$, $a_{499} = 0$ and for $n \geq 1$, $a_{n+1} = 2a_1 a_n - a_{n-1}$.

(a) Prove that $|a_1| \leq 1$.
(b) Find a_{1996}.

solution page 91

2

XII Italian Mathematical Olympiad, 1996

Prove that the equation $a^2 + b^2 = c^2 + 3$ has infinitely many integer solutions (a, b, c).

solution page 92

1

South African Mathematical Olympiad, 1995 (Section A)

Prove that there are no integers m and n such that

$$419m^2 + 95mn + 2000n^2 = 1995.$$

solution page 93

3

South African Mathematical Olympiad, 1995 (Section A)

Suppose that $a_1, a_2, a_3, \ldots, a_n$ are the numbers $1, 2, 3, \ldots, n$ but written in any order. Prove that

$$(a_1 - 1)^2 + (a_2 - 2)^2 + (a_3 - 3)^2 + \cdots + (a_n - n)^2$$

is always even.

solution page 94

2

South African Mathematical Olympiad, 1995 (Section B)

Find all pairs (m, n) of natural numbers with $m < n$ such that $m^2 + 1$ is a multiple of n and $n^2 + 1$ is a multiple of m.

solution page 95

2

Taiwan Mathematical Olympiad, 1996

Let a be a real number such that $0 < a \leq 1$ and $a \leq a_j \leq \frac{1}{a}$, for $j = 1, 2, \ldots, 1996$. Show that for any non-negative real numbers λ_j ($j = 1, 2, \ldots, 1996$), with

$$\sum_{j=1}^{1996} \lambda_j = 1,$$

one has

$$\left(\sum_{i=1}^{1996} \lambda_i a_i \right) \left(\sum_{j=1}^{1996} \lambda_j a_j^{-1} \right) \leq \frac{1}{4} \left(a + \frac{1}{a} \right)^2.$$

solution page 97

Taiwan Mathematical Olympiad, 1996

4

Show that for any real numbers $a_3, a_4, \ldots, a_{85}$, the roots of the equation

$$a_{85}x^{85} + a_{84}x^{84} + \cdots + a_3x^3 + 3x^2 + 2x + 1 = 0$$

are not real.

solution page 98

Croatian Mathematical Olympiad, 1995 (IV Class)

1

Is there any solution of the equation

$$\lfloor x \rfloor + \lfloor 2x \rfloor + \lfloor 4x \rfloor + \lfloor 8x \rfloor + \lfloor 16x \rfloor + \lfloor 32x \rfloor = 12345?$$

($\lfloor x \rfloor$ denotes the greatest integer which does not exceed x.)

solution page 99

Croatian Mathematical Olympiad, 1995 (IV Class)

2

Determine all pairs of numbers $\lambda_1, \lambda_2 \in \mathbb{R}$ for which every solution of the equation
$$(x + i\lambda_1)^n + (x + i\lambda_2)^n = 0$$
is real. Find the solutions.

solution page 100

Croatian Team Selection Test, 1995

3

Find all pairs of consecutive integers the difference of whose cubes is a full square.

solution page 101

13th Iranian Mathematical Olympiad, 1995 (Second Round)

1

Prove that for every natural number $n \geq 3$ there exist two sets $A = \{x_1, x_2, \ldots, x_n\}$ and $B = \{y_1, y_2, \ldots, y_n\}$ such that
(a) $A \cap B = \emptyset$,
(b) $x_1 + x_2 + \cdots + x_n = y_1 + y_2 + \cdots + y_n$,
(c) $x_1^2 + x_2^2 + \cdots + x_n^2 = y_1^2 + y_2^2 + \cdots + y_n^2$.

solution page 102

13th Iranian Mathematical Olympiad, 1995 (Second Round)

5

Prove that for any natural number n

$$\lceil \sqrt{n} + \sqrt{n+1} + \sqrt{n+2} \rceil = \lceil \sqrt{9n+8} \rceil.$$

solution page 103

13th Iranian Mathematical Olympiad, 1995 (Final Round)

1

Prove the following inequality

$$(xy + xz + yz)\left(\frac{1}{(x+y)^2} + \frac{1}{(y+z)^2} + \frac{1}{(x+z)^2}\right) \geq \frac{9}{4}$$

for positive real numbers x, y, z.

solution page 104

13th Iranian Mathematical Olympiad, 1995 (Final Round)

4

Let k be a positive integer. Prove that there are infinitely many perfect squares in the arithmetic progression $\{n \times 2^k - 7\}_{n \geq 1}$.

solution page 105

13th Iranian Mathematical Olympiad 1995

6

Let a, b, c be positive real numbers. Find all real numbers x, y, z such that

$$x + y + z = a + b + c$$

$$4xyz - (a^2 x + b^2 y + c^2 z) = abc.$$

solution page 106

Estonian Mathematical Contest, 1995 (Final Round)

1

The numbers x, y and $\dfrac{x^2 + y^2 + 6}{xy}$ are positive integers. Prove that

$\dfrac{x^2 + y^2 + 6}{xy}$ is a perfect cube.

solution page 107

Bi-National Israel-Hungary Competition, 1996

1

Find all sequences of integers x_1, x_2, ..., x_{1997} such that

$$\sum_{k=1}^{1997} 2^{k-1}(x_k)^{1997} = 1996 \prod_{k=1}^{1997} x_k.$$

solution page 108

Bi-National Israel-Hungary Competition, 1996

4

Let a_1, a_2, ..., a_n be arbitrary real numbers and b_1, b_2, ..., b_n real numbers satisfying the condition $1 \geq b_1 \geq b_2 \geq \cdots \geq b_n \geq 0$. Prove that there is a positive integer $k \leq n$ for which the inequality $|a_1 b_1 + a_2 b_2 + \cdots + a_n b_n| \leq |a_1 + a_2 + \cdots + a_k|$ holds.

solution page 109

Finnish High School Mathematics Contest, 1997 (Final Round)

1

Determine all numbers a, for which the equation

$$a3^x + 3^{-x} = 3$$

has a unique solution x.

solution page 110

Georgian Mathematical Olympiad, 1997 (X form)

1

Find all triples (x, y, z) of integers satisfying the inequality:

$$x^2 + y^2 + z^2 + 3 < xy + 3y + 2z.$$

solution page 111

6th ROC Taiwan Mathematical Olympiad, 1997 (Part I)

3

Let $n \geq 3$. Suppose that the sequence $a_1, a_2, \ldots, a_n$ of positive real numbers satisfies $a_{i-1} + a_{i+1} = k_i a_i$, $\forall i = 1, 2, \ldots, n$, where each k_i is a positive integer, $a_0 = a_n$, $a_{n+1} = a_1$. Show that

$$2n \leq k_1 + k_2 + \cdots + k_n \leq 3n.$$

solution page 112

11th Iberoamerican Mathematical Olympiad, 1996

4

(Brazil): Given a natural number $n \geq 2$, all the fractions of the form $\frac{1}{ab}$, with a and b natural numbers, coprime and such that

$$a < b \leq n, \qquad a + b > n,$$

are considered. Show that the sum of all these fractions equals $\frac{1}{2}$.

solution page 113

XXXIII Spanish Mathematical Olympiad, 1996

4

The sum of two of the roots of the equation

$$x^3 - 503x^2 + (a+4)x - a = 0$$

is equal to 4. Determine the value of a.

solution page 114

XXXIII Spanish Mathematical Olympiad, 1996

5

If a, b, c are positive real numbers, prove the inequality

$$a^2 + b^2 + c^2 - ab - bc - ca \geq 3(b-c)(a-b).$$

When is the "$=$" sign valid?

solution page 115

Chinese Mathematical Olympiad, 1997

6

Let $a_1, a_2, \ldots$ be non-negative numbers which satisfy

$$a_{n+m} \leq a_n + a_m \quad (m, n \in \mathbb{N}).$$

Prove that

$$a_n \leq ma_1 + \left(\frac{n}{m} - 1\right) a_m$$

for all $n \geq m$.

solution page 116

48th Polish Mathematical Competition, 1997 (Final Round)

1

The positive integers x_1, x_2, x_3, x_4, x_5, x_6, x_7 satisfy the conditions:

$$x_6 = 144 \quad \text{and} \quad x_{n+3} = x_{n+2}(x_{n+1} + x_n) \quad \text{for} \quad n = 1, 2, 3, 4.$$

Compute x_7.

solution page 117

48th Polish Mathematical Competition, 1997 (Final Round)

2

Solve the following system of equations in real numbers x, y, z:

$$3(x^2 + y^2 + z^2) = 1 \tag{1}$$
$$x^2y^2 + y^2z^2 + z^2x^2 = xyz(x + y + z)^3. \tag{2}$$

solution page 118

48th Polish Mathematical Competition, 1997 (Final Round)

4

The sequence $a_1, a_2, a_3, \ldots$ is defined by

$$a_1 = 0 \quad a_n = a_{\lfloor n/2 \rfloor} + (-1)^{n(n+1)/2} \quad \text{for} \quad n > 1.$$

For every integer $k \geq 0$, find the number of all n such that

$$2^k \leq n < 2^{k+1}, \quad a_n = 0$$

($\lfloor n/2 \rfloor$ denotes the greatest integer not exceeding $n/2$).

solution page 119

18th Brazilian Mathematical Olympiad, 1996

1

Show that the equation

$$x^2 + y^2 + z^2 = 3xyz$$

has infinitely many integer solutions with $x > 0$, $y > 0$ and $z > 0$.

solution page 120

For positive a and b and natural n prove

$$\frac{1}{a+b} + \frac{1}{a+2b} + \cdots + \frac{1}{a+nb} < \frac{n}{\sqrt{a(a+nb)}}.$$

solution page 121

Solve the system of equations in $\mathbb{R}^3$:

$$8(x^3 + y^3 + z^3) = 73,$$
$$2(x^2 + y^2 + z^2) = 3(xy + yz + zx),$$
$$xyz = 1.$$

solution page 122

Let n and r be positive integers such that $n \geq 2$ and $r \not\equiv 0$ (mod n), and let g be the greatest common divisor of n and r. Prove that

$$\sum_{i=1}^{n-1} \left\langle \frac{ri}{n} \right\rangle = \frac{1}{2}(n - g),$$

where $\langle x \rangle = x - \lfloor x \rfloor$ is the fractional part of x.

solution page 123

Let k and n be integers such that $1 \leq k \leq n$, and assume that a_1, $a_2, \ldots, a_k$ satisfy

$$a_1 + a_2 + \cdots + a_k = n,$$
$$a_1^2 + a_2^2 + \cdots + a_k^2 = n,$$
$$\cdots\cdots\cdots\cdots\cdots$$
$$a_1^k + a_2^k + \cdots + a_k^k = n.$$

Prove that

$$(x + a_1)(x + a_2)\cdots(x + a_k) = x^k + \binom{n}{1}x^{k-1} + \binom{n}{2}x^{k-2} + \cdots + \binom{n}{k}.$$

solution page 125

Solve, in integers, the equation

$$(x^2 - y^2)^2 = 1 + 16y.$$

solution page 126

Given all possible quadratic trinomials of the type $x^2 + px + q$, with integer coefficients p and q, $1 \leq p \leq 1997$, $1 \leq q \leq 1997$. Consider the sets of the trinomials:

(a) having integer zeros,

(b) not having real zeros.

Which of those sets is larger?

solution page 128

(a) Prove that for all real numbers p and q the inequality $p^2 + q^2 + 1 > p(q+1)$ holds.

(b) Determine the greatest real number b such that for all real numbers p and q the inequality $p^2 + q^2 + 1 > bp(q+1)$ holds.

(c) Determine the greatest real number c such that for all integers p and q the inequality $p^2 + q^2 + 1 > cp(q+1)$ holds.

solution page 130

Find all real solutions of

$$\sqrt[4]{13 + x} + \sqrt[4]{4 - x} = 3.$$

solution page 131

Find

(a) all quadruples of positive integers (a, k, l, m) for which the equality $a^k = a^l + a^m$ holds;

(b) all 5-tuples of positive integers (a, k, l, m, n) for which the equality $a^k = a^l + a^m + a^n$ holds.

solution page 132

Prove that, for any real numbers x and y, the following inequality holds:

$$x^2 + y^2 + 1 > x\sqrt{y^2 + 1} + y\sqrt{x^2 + 1}.$$

solution page 133

Estonian Mathematical Olympiad, 1997 (Final Round)

6

For positive integers m, n denote $T(m, n) = \gcd\left(m, \frac{n}{\gcd(m,n)}\right)$.

(a) Prove that there exist infinitely many pairs of integers (m, n) such that $T(m, n) > 1$ and $T(n, m) > 1$.

(b) Does there exist a pair of integers (m, n) such that $T(m, n) = T(n, m) > 1$?

solution page 134

Ukrainian Mathematical Olympiad, 1997

2

Solve the system in real numbers

$$\begin{cases} x_1 + x_2 + \cdots + x_{1997} = 1997 \\ x_1^4 + x_2^4 + \cdots + x_{1997}^4 = x_1^3 + x_2^3 + \cdots + x_{1997}^3 . \end{cases}$$

solution page 135

Ukrainian Mathematical Olympiad, 1997

5

It is known that the equation $ax^3 + bx^2 + cx + d = 0$ with respect to x has three distinct real roots. How many roots does the equation $4(ax^3 + bx^2 + cx + d)(3ax + b) = (3ax^2 + 2bx + c)^2$ have?

solution page 136

10th Irish Mathematical Olympiad, 1997

1

Find (with proof) all pairs of integers (x, y) satisfying the equation

$$1 + 1996x + 1998y = xy.$$

solution page 137

10th Irish Mathematical Olympiad, 1997

4

Let a, b, c be non-negative real numbers such that $a + b + c \geq abc$. Prove that $a^2 + b^2 + c^2 \geq abc$.

solution page 139

10th Irish Mathematical Olympiad, 1997

Let S be the set of all odd integers greater than one. For each $x \in S$, denote by $\delta(x)$ the unique integer satisfying the inequality

$$2^{\delta(x)} < x < 2^{\delta(x)+1} .$$

For $a, b \in S$, define

$$a * b = 2^{\delta(a)-1}(b - 3) + a .$$

[For example, to calculate $5 * 7$ note that $2^2 < 5 < 2^3$, so that $\delta(5) = 2$, and hence, $5 * 7 = 2^{2-1}(7 - 3) + 5 = 13$. Also $2^2 < 7 < 2^3$, so that $\delta(7) = 2$ and $7 * 5 = 2^{2-1}(5 - 3) + 7 = 11$].

Prove that if $a, b, c \in S$, then

(a) $a * b \in S$ and

(b) $(a * b) * c = a * (b * c)$.

solution page 141

Hungary-Israel Bi-National Mathematical Competition, 1997

Is there an integer N such that

$$(\sqrt{1997} - \sqrt{1996})^{1998} = \sqrt{N} - \sqrt{N - 1}?$$

solution page 142

Hubgary-Israel Bi-National Mathematical Competition, 1997

Find all real numbers α with the following property: for any positive integer n there exists an integer m such that

$$\left| \alpha - \frac{m}{n} \right| < \frac{1}{3n} .$$

solution page 143

St. Petersburg City Mathematical Olympiad, 1997 (10th Grade)

Positive integers x, y, z satisfy the equation $2x^x + y^y = 3z^z$. Prove that they are equal.

solution page 144

St. Petersburg City Mathematical Olympiad, 1997 (11th Grade)

Prove that for $x \geq 2, y \geq 2, z \geq 2$

$$(y^3 + x)(z^3 + y)(x^3 + z) \geq 125xyz .$$

solution page 145

Vietnamese Team Selection Test, 1997

Find the greatest real number α such that there exists an infinite sequence of whole numbers (a_n) $(n = 1, 2, 3, \dots)$ satisfying simultaneously the following conditions:

(i) $a_n > 1997^n$ for every $n \in \mathbb{N}^*$,

(ii) $a_n^\alpha \leq U_n$ for every $n \geq 2$, where U_n is the greatest common divisor of the set of numbers $\{a_i + a_j \mid i + j = n\}$.

solution page 146

Vietnamese Team Selection Test, 1997

Determine all pairs of positive real numbers a, b such that for every $n \in \mathbb{N}^*$ and for every real root x_n of the equation

$$4n^2 x = \log_2(2n^2 x + 1)$$

we have

$$a^{x_n} + b^{x_n} \geq 2 + 3x_n .$$

solution page 148

7th Japan Mathematical Olympiad, 1997 (Final Round)

Let a, b, c be positive integers. Prove that the inequality

$$\frac{(b+c-a)^2}{(b+c)^2 + a^2} + \frac{(c+a-b)^2}{(c+a)^2 + b^2} + \frac{(a+b-c)^2}{(a+b)^2 + c^2} \geq \frac{3}{5}$$

holds. Determine also when the equality holds.

solution page 149

Vietnamese Mathematical Olympiad, 1997

Let there be given a whole number $n > 1$, not divisible by 1997. Consider two sequences of numbers $\{a_i\}$ and $\{b_j\}$ defined by:

$$a_i = i + \frac{ni}{1997} \quad (i = 1, 2, 3, \dots, 1996),$$

$$b_j = j + \frac{1997j}{n} \quad (j = 1, 2, 3, \dots, n-1).$$

By arranging the numbers of these two sequences in increasing order, we get the sequence $c_1 \leq c_2 \leq \cdots \leq c_{1995+n}$.

Prove that $c_{k+1} - c_k < 2$ for every $k = 1, 2, \dots, 1994 + n$.

solution page 150

38 IMO Turkey Team Selection Test, 1997

Prove that, for each prime number $p \geq 7$, there exists a positive integer n and integers $x_1, x_2, \ldots, x_n, y_1, y_2, \ldots, y_n$ which are not divisible by p, such that

$$
\begin{aligned}
x_1^2 + y_1^2 &\equiv x_2^2 \pmod{p}, \\
x_2^2 + y_2^2 &\equiv x_3^2 \pmod{p}, \\
&\;\;\vdots \\
x_{n-1}^2 + y_{n-1}^2 &\equiv x_n^2 \pmod{p}, \\
x_n^2 + y_n^2 &\equiv x_1^2 \pmod{p}.
\end{aligned}
$$

solution page 152

38 IMO Turkey Team Selection Test, 1997

Given an integer $n \geq 2$, find the minimal value of

$$
\frac{x_1^5}{x_2 + x_3 + \cdots + x_n} + \frac{x_2^5}{x_1 + x_3 + \cdots + x_n} + \cdots + \frac{x_n^5}{x_1 + x_2 + \cdots + x_{n-1}}
$$

subject to $x_1^2 + x_2^2 + \cdots + x_n^2 = 1$, where $x_1, x_2, \ldots, x_n$ are positive real numbers.

solution page 155

Chilean Mathematical Olympiad, 1994

Let x be a number such that

$$
x + \frac{1}{x} = -1.
$$

Compute

$$
x^{1994} + \frac{-1}{x^{1994}}.
$$

solution page 155

Chilean Mathematical Olympiad, 1994

Let a be a natural number. Show that the equation

$$
x^2 - y^2 = a^3
$$

always has integer solutions for x and y.

solution page 156

24-th Spanish Olympiad

Solve the following system of equations in the set of complex numbers:
$$|z_1| = |z_2| = |z_3| = 1,$$
$$z_1 + z_2 + z_3 = 1,$$
$$z_1 z_2 z_3 = 1.$$

solution page 157

IMO Proposal by Poland (1985)

Given nonnegative real numbers $x_1, x_2, \ldots, x_k$ and positive integers k, m, n such that $km \leq n$, prove that

$$n \left(\prod_{i=1}^{k} x_i^m - 1 \right) \leq m \sum_{i=1}^{k} (x_i^n - 1).$$

solution page 158

Bulgarian Mathematical Olympiad 1984

Let $0 \leq x_i \leq 1$ and $x_i + y_i = 1$, for $i = 1, 2, \ldots, n$. Prove that

$$(1 - x_1 x_2 \cdots x_n)^m + (1 - y_1^m)(1 - y_2^m) \cdots (1 - y_n^m) \geq 1$$

for all positive integers m and n.

solution page 159

Annual Greek High School Competition 1983

If $\mathbf{a}$ and $\mathbf{b}$ are given nonparallel vectors, solve for x in the equation

$$\frac{\mathbf{a}^2 + x\mathbf{a} \cdot \mathbf{b}}{|\mathbf{a}||\mathbf{a} + x\mathbf{b}|} = \frac{\mathbf{b}^2 + \mathbf{a} \cdot \mathbf{b}}{|\mathbf{b}||\mathbf{a} + \mathbf{b}|}.$$

solution page 160

IMO Proposal by Canada (1985)

Prove that

$$\frac{x_1^2}{x_1^2 + x_2 x_3} + \frac{x_2^2}{x_2^2 + x_3 x_1} + \cdots + \frac{x_{n-1}^2}{x_{n-1}^2 + x_n x_1} + \frac{x_n^2}{x_n^2 + x_1 x_2} \leq n - 1,$$

where all $x_i > 0$.

solution page 161

19th Austrian Mathematical Olympiad, Final Round

Let $a_1, \ldots, a_{1988}$ be positive real numbers whose arithmetic mean equals 1988. Show

$$\sqrt[1988]{\prod_{i=1}^{1988} \prod_{j=1}^{1988} (1 + \frac{a_i}{a_j})} \geq 2^{1988}$$

and determine when equality holds.

solution page 162

US Olympiad Student Proposals (1982)

4

Find all solutions (x, y, z) of the Diophantine equation

$$x^3 + y^3 + z^3 + 6xyz = 0.$$

solution page 163

US Olympiad Student Proposals (1982)

7

In $\mathbf{R}^n$ let $\mathbf{X} = (x_1, x_2, \ldots, x_n)$, $\mathbf{Y} = (y_1, y_2, \ldots, y_n)$, and, for $p \in (0, 1)$, define

$$F_p(\mathbf{X}, \mathbf{Y}) \equiv \left(\left| \frac{x_1}{p} \right|^p \left| \frac{y_1}{1-p} \right|^{1-p}, \left| \frac{x_2}{p} \right|^p \left| \frac{y_2}{1-p} \right|^{1-p}, \ldots, \left| \frac{x_n}{p} \right|^p \left| \frac{y_n}{1-p} \right|^{1-p} \right).$$

Prove that

$$\|\mathbf{X}\|_m + \|\mathbf{Y}\|_m \geq \|F_p(\mathbf{X}, \mathbf{Y})\|_m,$$

where

$$\|\mathbf{X}\|_m = (|x_1|^m + |x_2|^m + \cdots + |x_n|^m)^{1/m}.$$

solution page 164

Bulgarian Winter Competition 1983

3

Determine all values of the real parameter p for which the system of equations

$$x + y + z = 2$$
$$yz + zx + xy = 1$$
$$xyz = p$$

has a real solution.

solution page 165

British Mathematical Olympiad 1983

3

The real numbers $x_1, x_2, x_3, \ldots$ are defined by

$$x_1 = a \neq -1 \quad \text{and} \quad x_{n+1} = x_n^2 + x_n \text{ for all } n \geq 1.$$

S_n is the sum and P_n is the product of the first n terms of the sequence $y_1, y_2, y_3, \ldots$, where

$$y_n = \frac{1}{1 + x_n}.$$

Prove that $aS_n + P_n = 1$ for all n.

solution page 166

Asian Pacific Mathematical Olympiad 1989

1

Let $x_1, x_2, \ldots, x_n$ be positive real numbers, and let $S = x_1 + x_2 + \cdots + x_n$. Prove that

$$(1 + x_1)(1 + x_2) \cdots (1 + x_n) \leq 1 + S + \frac{S}{2!} + \frac{S}{3!} + \cdots + \frac{S}{n!}.$$

solution page 167

Austrian-Polish Mathematics Competition 1982

3

Prove that, for all natural numbers $n \geq 2$,

$$\prod_{i=1}^{n} \tan\left\{\frac{\pi}{3}\left(1 + \frac{3^i}{3^n - 1}\right)\right\} = \prod_{i=1}^{n} \cot\left\{\frac{\pi}{3}\left(1 - \frac{3^i}{3^n - 1}\right)\right\} .$$

solution page 168

Austrian-Polish Mathematics Competition 1985

4

Determine all real solutions x, y of the system

$$x^4 + y^2 - xy^3 - 9x/8 = 0.$$
$$y^4 + x^2 - yx^3 - 9y/8 = 0.$$

solution page 169

Leningrad High School Olympiad (Third Round) 1982

8

Prove that for any natural number k, there is an integer n such that

$$\sqrt{n + 1981^k} + \sqrt{n} = (\sqrt{1982} + 1)^k .$$

solution page 171

3rd Ibero-American Olympiad

2

Let a, b, c, d, p and q be natural numbers different from zero such that

$$ad - bc = 1 \quad \text{and} \quad \frac{a}{b} > \frac{p}{q} > \frac{c}{d} .$$

Show that
  (i) $q \geq b + d$;
  (ii) if $q = b + d$ then $p = a + c$.

solution page 172

3rd Ibero-American Olympiad

5

We consider expressions of the form

$$x + yt + zt^2 ,$$

where $x, y, z \in \mathbf{Q}$, and $t^2 = 2$. Show that, if $x + yt + zt^2 \neq 0$, then there exist $u, v, w \in \mathbf{Q}$ such that

$$(x + yt + zt^2)(u + vt + wt^2) = 1 .$$

solution page 173

Spanish Mathematical Olympiad, 1st Round 1985

5

Determine all the real roots of $4x^4 + 16x^3 - 6x^2 - 40x + 25 = 0$.

solution page 174

Spanish Mathematical Olympiad, 2nd Round 1985

7

Determine the value of p such that the equation $x^5 - px - 1 = 0$ has two roots r and s which are the roots of an equation $x^2 - ax + b = 0$ where a and b are integers.

solution page 175

IMO Proposal by Ireland (1989)

9

Let a, b, c, d, m, n be positive integers such that

$$a^2 + b^2 + c^2 + d^2 = 1989.$$
$$a + b + c + d = m^2.$$

and the largest of a, b, c, d is n^2. Determine, with proof, the values of m and n.

solution page 176

Canadian Mathematics Olympiad 1991

1

Show that the equation $x^2 + y^5 = z^3$ has infinitely many solutions in integers x, y, z for which $xyz \neq 0$.

solution page 177

Asian Pacific Mathematical Olympiad 1989

2

Prove that the equation

$$6(6a^2 + 3b^2 + c^2) = 5n^2$$

has no solution in integers except $a = b = c = n = 0$.

solution page 178

USAMO Training Session 1986

2

Determine the maximum value of

$$x^3 + y^3 + z^3 - x^2y - y^2z - z^2x$$

for $0 \leq x, y, z \leq 1$.

solution page 179

Austrian-Polish Mathematical Competition 1986

5

Determine all quadruples (x, y, u, v) of real numbers satisfying the simultaneous equations

$$x^2 + y^2 + u^2 + v^2 = 4,$$
$$xu + yv = -xv - yu,$$
$$xyu + yuv + uvx + vxy = -2,$$
$$xyuv = -1.$$

solution page 180

Singapore MSI Mathematical Competition 1988

5

Find all positive integers x, y, z satisfying the equation $5(xy + yz + zx) = 4xyz$.

solution page 181

Tenth Atlantic Provinces Mathematics Competition

8

Find the sum of the infinite series

$$1 + \frac{1}{2} + \frac{1}{3} + \frac{1}{4} + \frac{1}{6} + \frac{1}{8} + \frac{1}{9} + \frac{1}{12} + \cdots$$

where the terms are reciprocals of integers divisible only by the primes 2 or 3.

solution page 182

29th I.M.O. in Australia

6

Let a and b be positive integers such that $ab + 1$ divides $a^2 + b^2$. Show that

$$\frac{a^2 + b^2}{ab + 1}$$

is the square of an integer.

solution page 183

XIV ALL UNION Mathematical Olympiad U.S.S.R.

5

Does the equation $x^2 + y^3 = z^4$ have solutions for prime numbers x, y and z?

solution page 184

43rd Mathematical Olympiad (1991-92) in Poland (Final round)

3

Prove that the inequality

$$\sum_{n=1}^{r} \left(\sum_{m=1}^{r} \frac{a_m a_n}{m + n} \right) \geq 0$$

holds for any real numbers $a_1, a_2, \ldots, a_r$. Find conditions for equality.

solution page 185

34th IMO Proposal by USA (1993)

Prove that

$$\frac{a}{b+2c+3d} + \frac{b}{c+2d+3a} + \frac{c}{d+2a+3b} + \frac{d}{a+2b+3c} \geq \frac{2}{3}$$

for all positive real numbers a, b, c, d.

solution page 186

Nordic Mathematical Contest, 1992

Determine all real numbers x, y, z greater than 1, satisfying the equation

$$x + y + z + \frac{3}{x-1} + \frac{3}{y-1} + \frac{3}{z-1} = 2\left(\sqrt{x+2} + \sqrt{y+2} + \sqrt{z+2}\right).$$

solution page 187

10th Iranian Mathematical Olympiad

Find all integer solutions of

$$\frac{1}{m} + \frac{1}{n} - \frac{1}{mn^2} = \frac{3}{4}.$$

solution page 188

10th Iranian Mathematical Olympiad

Let a, b, c, be rational and one of the roots of $ax^3 + bx + c = 0$ be equal to the product of the other two roots. Prove that this root is rational.

solution page 189

Canadian Mathematical Olympiad 1996

If α, β, γ are the roots of $x^3 - x - 1 = 0$, compute

$$\frac{1+\alpha}{1-\alpha} + \frac{1+\beta}{1-\beta} + \frac{1+\gamma}{1-\gamma}.$$

solution page 190

Canadian Mathematical Olympiad 1996

Find all real solutions to the following system of equations. Carefully justify your answer.

$$\begin{cases} \dfrac{4x^2}{1+4x^2} = y \\[2ex] \dfrac{4y^2}{1+4y^2} = z \\[2ex] \dfrac{4z^2}{1+4z^2} = x \end{cases}$$

solution page 191

10th Iranian Mathematical Olympiad (Second Stage Exam)

2

Given the sequence $a_0 = 1$, $a_1 = 2$, $a_{n+1} = a_n + \frac{a_{n-1}}{1+(a_{n-1})^2}$, $n > 1$, show that $52 < a_{1371} < 65$.

solution page 192

Turkish Mathematical Olympiad, 1993 (Final Selection Test)

3

Let $\{b_n\}$ be sequence of positive real numbers such that

for each $n \geq 1$, $\quad b_{n+1}^2 \geq \frac{b_1^2}{1^3} + \frac{b_2^2}{2^3} + \cdots + \frac{b_n^2}{n^3}$.

Show that there is a natural number K such that

$$\sum_{n=1}^{K} \frac{b_{n+1}}{b_1 + b_2 + \cdots + b_n} > \frac{1993}{1000}.$$

solution page 193

6th Irish Mathematical Olympiad, 1993

1

The real numbers α, β satisfy the equations

$\alpha^3 - 3\alpha^2 + 5\alpha - 17 = 0$. $\qquad \beta^3 - 3\beta^2 + 5\beta + 11 = 0$.

Find $\alpha + \beta$.

solution page 194

6th Irish Mathematical Olympiad, 1993

7

For non-negative integers n, r the binomial coefficient $\binom{n}{r}$ denotes the number of combinations of n objects chosen r at a time, with the convention that $\binom{n}{0} = 1$ and $\binom{n}{r} = 0$ if $n < r$. Prove the identity

$$\sum_{d=1}^{\infty} \binom{n-r+1}{d}\binom{r-1}{d-1} = \binom{n}{r}$$

for all integers n, r with $1 \leq r \leq n$.

solution page 195

Dutch Mathematical Olympiad, 1992 (Second Round)

4

For every positive integer n, $n?$ is defined as follows:

$$n? = \begin{cases} 1 & for \ n = 1 \\ \frac{n}{(n-1)?} & for \ n \geq 2 \end{cases}$$

Prove $\sqrt{1992} < 1992? < \frac{4}{3}\sqrt{1992}$.

solution page 196

16th Austrian Polish Mathematics Competition, 1993 (First Day) [1]

Determine all natural numbers $x, y \geq 1$ such that $2^x - 3^y = 7$.

solution page 197

16th Austrian Polish Mathematics Competition, 1993 (First Day) [5]

Determine all real solutions x, y, z of the system of equations:

$$x^3 + y = 3x + 4,$$
$$2y^3 + z = 6y + 6,$$
$$3z^3 + x = 9z + 8.$$

solution page 198

16th Austrian Polish Mathematics Competition, 1993 (First Day) [6]

Show: For all real numbers $a, b \geq 0$ the following chain of inequalities is valid

$$\left(\frac{\sqrt{a} + \sqrt{b}}{2}\right)^2 \leq \frac{a + \sqrt[3]{a^2 b} + \sqrt[3]{ab^2} + b}{4}$$

$$\leq \frac{a + \sqrt{ab} + b}{3} \leq \sqrt{\left(\frac{\sqrt[3]{a^2} + \sqrt[3]{b^2}}{2}\right)^3}.$$

Also, for all three inequalities determine the cases of equality.

solution page 199

32nd Ukrainian Mathematical Olympiad, 1992 [2]

There are real numbers a, b, c, such that $a \geq b \geq c > 0$. Prove that

$$\frac{a^2 - b^2}{c} + \frac{c^2 - b^2}{a} + \frac{a^2 - c^2}{b} \geq 3a - 4b + c.$$

solution page 200

32nd Ukrainian Mathematical Olympiad, 1992 [5]

Prove that there are no real numbers x, y, z, such that

$$x^2 + 4yz + 2z = 0,$$
$$x + 2xy + 2z^2 = 0,$$
$$2xz + y^2 + y + 1 = 0.$$

solution page 201

36th International Mathematical Olympiad, 1995 (Canada) [2]

Let a, b, and c be positive real numbers such that $abc = 1$. Prove that

$$\frac{1}{a^3(b+c)} + \frac{1}{b^3(c+a)} + \frac{1}{c^3(a+b)} \geq \frac{3}{2}.$$

solution page 202

Prove that

$$\frac{1}{1999} < \frac{1}{2} \cdot \frac{3}{4} \cdot \frac{5}{6} \cdots \frac{1997}{1998} < \frac{1}{44}.$$

solution page 203

Let n be a positive integer, $a_1, a_2, \ldots, a_n$ positive real numbers and $s = a_1 + a_2 + \cdots + a_n$. Prove that

$$\sum_{i=1}^{n} \frac{a_i}{s - a_i} \geq \frac{n}{n-1} \qquad \text{and} \qquad \sum_{i=1}^{n} \frac{s - a_i}{a_i} \geq n(n-1).$$

solution page 204

Show that if x, y, z are positive integers such that $x^2 + y^2 + z^2 = 1993$, then $x + y + z$ is not a perfect square.

solution page 205

Given that $a^2 + b^2 + (a + b)^2 = c^2 + d^2 + (c + d)^2$, prove that $a^4 + b^4 + (a + b)^4 = c^4 + d^4 + (c + d)^4$.

solution page 206

Prove that the product of the 99 numbers $\frac{k^3-1}{k^3+1}$, $k = 2, 3, \ldots, 100$, is greater than $\frac{2}{3}$.

solution page 207

Find all integers satisfying the equation

$$2^x \cdot (4 - x) = 2x + 4.$$

solution page 208

Prove that for any positive $x_1, x_2, \ldots, x_n, y_1, y_2, \ldots, y_n$ the inequality

$$\sum_{i=1}^{n} \frac{1}{x_i y_i} \geq \frac{4n^2}{\sum_{i=1}^{n}(x_i + y_i)^2}$$

solution page 209

holds.

6th Irish Mathematical Olympiad

3

For nonnegative integers n, r the binomial coefficient $\binom{n}{r}$ denotes the number of combinations of n objects chosen r at a time, with the convention that $\binom{n}{0} = 1$ and $\binom{n}{r} = 0$ if $n < r$. Prove the identity

$$\sum_{d=1}^{\infty} \binom{n-r+1}{d}\binom{r-1}{d-1} = \binom{n}{r}$$

for all integers n, r with $1 \leq r \leq n$.

solution page 210

6th Irish Mathematical Olympiad

4

Let x be a real number with $0 < x < \pi$. Prove that, for all natural numbers n, the sum

$$\sin x + \frac{\sin 3x}{3} + \frac{\sin 5x}{5} + \cdots + \frac{\sin(2n-1)x}{2n-1}$$

is positive.

solution page 211

44th Latvian Mathematical Olympiad, 1994 (Final Grade, 3rd Round)

1

It is given that $\cos x = \cos y$ and $\sin x = -\sin y$. Prove that $\sin 1994x + \sin 1994y = 0$.

solution page 212

44th Latvian Mathematical Olympiad, 1994 (Final Grade, 3rd Round)

3

It is given that $a > 0, b > 0, c > 0, a + b + c = abc$. Prove that at least one of the numbers a, b, c exceeds $17/10$.

solution page 213

44th Latvian Mathematical Olympiad, 1994 (Final Grade, 3rd Round)

4

Solve the equation $1! + 2! + 3! + \cdots + n! = m^3$ in natural numbers.

solution page 214

44th Latvian Mathematical Olympiad, 1994 (1st Selection Test)

1

It is given that x and y are positive integers and $3x^2 + x = 4y^2 + y$. Prove that $x - y$, $3x + 3y + 1$ and $4x + 4y + 1$ are squares of integers.

solution page 215

44th Latvian Mathematical Olympiad, 1994 (2nd Selection Test)

1

It is given that $0 \leq x_i \leq 1$, $i = 1, 2, \ldots, n$. Find the maximum of the expression

$$\frac{x_1}{x_2 x_3 \ldots x_n + 1} + \frac{x_2}{x_1 x_3 x_4 \ldots x_n + 1} + \cdots + \frac{x_n}{x_1 x_2 \ldots x_{n-1} + 1}.$$

solution page 216

44th Latvian Mathematical Olympiad, 1994

3

It is given that $a > 0$, $b > 0$, $c > 0$, $a + b + c = abc$. Prove that at least one of the numbers a, b, c exceeds $17/10$.

solution page 217

Bi-National Israel-Hungary Mathematical Competition, 1994

1

$a_1, \ldots, a_k, a_{k+1}, \ldots, a_n$ are positive numbers ($k < n$). Suppose that the values of $a_{k+1}, \ldots, a_n$ are fixed. How should one choose the values of $a_1, \ldots, a_n$ in order to minimize $\sum_{i,j, i \neq j} \frac{a_i}{a_j}$?

solution page 218

Bi-National Israel-Hungary Mathematical Competition, 1994

3

m, n are 2 different natural numbers. Show that there exists a real number x, such that $\frac{1}{3} \leq \{xn\} \leq \frac{2}{3}$ and $\frac{1}{3} \leq \{xm\} \leq \frac{2}{3}$, where $\{a\}$ is the fractional part of a.

solution page 219

Canadian Mathematical Olympiad, 1998

1

Determine the number of real solutions a to the equation

$$\left[\frac{1}{2} a \right] + \left[\frac{1}{3} a \right] + \left[\frac{1}{5} a \right] = a.$$

Here, if x is a real number, then $[x]$ denotes the greatest integer that is less than or equal to x.

solution page 220

Canadian Mathematical Olympiad, 1998

2

Find all real numbers x such that

$$x = \left(x - \frac{1}{x} \right)^{1/2} + \left(1 - \frac{1}{x} \right)^{1/2}.$$

solution page 221

Canadian Mathematical Olympiad, 1998

3

Let n be a natural number such that $n \geq 2$. Show that

$$\frac{1}{n+1} \left(1 + \frac{1}{3} + \cdots + \frac{1}{2n-1} \right) > \frac{1}{n} \left(\frac{1}{2} + \frac{1}{4} + \cdots + \frac{1}{2n} \right).$$

solution page 222

Swedish Mathematical Olympiad, 1993

3

Assume that a and b are integers. Prove that the equation $a^2 + b^2 + x^2 = y^2$ has an integer solution x, y if and only if the product ab is even.

solution page 223

Swedish Mathematical Olympiad, 1993

4

To each pair of real numbers a and b, where $a \neq 0$ and $b \neq 0$, there is a real number $a * b$ such that

$$a * (b * c) = (a * b) \cdot c,$$
$$a * a = 1.$$

Solve the equation $x * 36 = 216$.

solution page 224

Hong Kong Committee - Mock Test, Part 2, IMO 1994

1

Suppose that $yz + zx + xy = 1$ and x, y, and $z \geq 0$. Prove that

$$x(1 - y^2)(1 - z^2) + y(1 - z^2)(1 - x^2) + z(1 - x^2)(1 - y^2) \leq \frac{4\sqrt{3}}{9}.$$

solution page 225

45th Mathematical Olympiad in Poland, 1994 (Final Round)

1

Determine all triples of positive rational numbers (x, y, z) such that $x + y + z$, $x^{-1} + y^{-1} + z^{-1}$ and xyz are integers.

solution page 226

45th Mathematical Olympiad in Poland, 1994 (Final Round)

5

Let $A_1, A_2, \ldots, A_8$ be the vertices of a parallelepiped and let O be its centre. Show that

$$4(OA_1^2 + OA_2^2 + \cdots + OA_8^2) \leq (OA_1 + OA_2 + \cdots + OA_8)^2.$$

solution page 227

Irish Mathematical Olympiad, 1994

1

Let x, y be positive integers with $y > 3$ and

$$x^2 + y^4 = 2[(x - 6)^2 + (y + 1)^2].$$

Prove that $x^2 + y^4 = 1994$.

solution page 228

Irish Mathematical Olympiad, 1994

6

A sequence $\{x_n\}$ is defined by the rules

$$x_1 = 2$$

and

$$nx_n = 2(2n - 1)x_{n-1}; \qquad n = 2, 3, \ldots.$$

Prove that x_n is an integer for every positive integer n.

solution page 229

Irish Mathematical Olympiad, 1994

Let p, q, r be distinct real numbers which satisfy the equations

$$q = p(4 - p),$$
$$r = q(4 - q),$$
$$p = r(4 - r).$$

Find all possible values of $p + q + r$.

solution page 250

Irish Mathematical Olympiad, 1994

Let w, a, b, c be distinct real numbers with the property that there exist real numbers x, y, z for which the following equations hold:

$$
\begin{aligned}
x + y + z &= 1 \\
xa^2 + yb^2 + zc^2 &= w^2 \\
xa^3 + yb^3 + zc^3 &= w^3 \\
xa^4 + yb^4 + zc^4 &= w^4.
\end{aligned}
$$

Express w in terms of a, b, c.

solution page 251

37th International Mathematical Olympiad, 1996 (Shortlist)

Let a, b and c be positive real numbers such that $abc = 1$. Prove that

$$\frac{ab}{a^5 + b^5 + ab} + \frac{bc}{b^5 + c^5 + bc} + \frac{ca}{c^5 + a^5 + ca} \leq 1.$$

When does equality hold?

solution page 252

Croatian National Mathematics Competition, 1994

Find all ordered triples (a, b, c) of real numbers such that for every three integers x, y, z the following identity holds:

$$|ax + by + cz| + |bx + cy + az| + |cx + ay + bz| = |x| + |y| + |z|.$$

solution page 253

17th Austrian-Polish Mathematics Competition, 1994

Solve the equation

$$\frac{1}{2}(x + y)(y + z)(z + x) + (x + y + z)^3 = 1 - xyz$$

in integers.

solution page 254

Let $n > 1$ be an odd positive integer. Assume that the integers x_1, $x_2, \ldots, x_n \geq 0$ satisfy the system of equations

$$\begin{aligned}
(x_2 - x_1)^2 + 2(x_2 + x_1) + 1 &= n^2, \\
(x_3 - x_2)^2 + 2(x_3 + x_2) + 1 &= n^2, \\
\cdots\cdots\cdots\cdots\cdots\cdots\cdots\cdots \\
(x_1 - x_n)^2 + 2(x_1 + x_n) + 1 &= n^2.
\end{aligned}$$

Show that either $x_1 = x_n$, or there exists j with $1 \leq j \leq n - 1$, such that $x_j = x_{j+1}$.

solution page 255

Let n and r be natural numbers. Find the smallest natural number m satisfying this condition: For each partition of the set $\{1, 2, \ldots, m\}$ into r subsets $A_1, A_2, \ldots, A_r$, there exist two numbers a and b in some A_i $(1 \leq i \leq r)$ such that $1 < \frac{a}{b} \leq 1 + \frac{1}{n}$.

solution page 256

Solve the following system of equations:

$$x \cdot |x| + y \cdot |y| = 1, \qquad \lfloor x \rfloor + \lfloor y \rfloor = 1,$$

in which $|t|$ and $\lfloor t \rfloor$ represent the absolute value and the integer part of the real number t.

solution page 257

Let $a \in \mathbb{R}$ be given. Find the real numbers $x_1, \ldots, x_n$ which satisfy the system of equations

$$\begin{aligned}
x_1^2 + ax_1 + \left(\tfrac{a-1}{2}\right)^2 &= x_2, \\
x_2^2 + ax_2 + \left(\tfrac{a-1}{2}\right)^2 &= x_3, \\
\cdots\cdots\cdots\cdots\cdots\cdots\cdots \\
x_{n-1}^2 + ax_{n-1} + \left(\tfrac{a-1}{2}\right)^2 &= x_n, \\
x_n^2 + ax_n + \left(\tfrac{a-1}{2}\right)^2 &= x_1.
\end{aligned}$$

solution page 258

Show that $\cos(\sin x) > \sin(\cos x)$ holds for every real number x.

solution page 259

44th Lithuanian Mathematical Olympiad (Grade XII)

1

Consider all pairs of real numbers satisfying the inequalities

$$-1 \leq x+y \leq 1, \qquad -1 \leq xy+x+y \leq 1.$$

Let M denote the largest possible value of x.

(a) Prove that $M \leq 3$.

(b) Prove that $M \leq 2$.

(c) Find M.

solution page 240

Canadian Mathematical Olympiad, 1999

1

Find all real solutions to the equation $4x^2 - 40[x] + 51 = 0$. Here, if x is a real number, then $[x]$ denotes the greatest integer that is less than or equal to x.

solution page 241

Canadian Mathematical Olympiad, 1999

4

Suppose a_1, a_2, ..., a_8 are eight distinct integers from $\{1, 2, \ldots, 16, 17\}$. Show that there is an integer $k > 0$ such that the equation $a_i - a_j = k$ has at least three different solutions. Also, find a specific set of 7 distinct integers from $\{1, 2, \ldots, 16, 17\}$ such that the equation $a_i - a_j = k$ does not have three distinct solutions for any $k > 0$.

solution page 242

Canadian Mathematical Olympiad, 1999

5

Let x, y, and z be non-negative real numbers satisfying $x + y + z = 1$. Show that

$$x^2 y + y^2 z + z^2 x \leq \frac{4}{27},$$

and find when equality occurs.

solution page 243

8 th Korean Mathematical Olympiad (First Round)

2

For a given positive integer m, find all pairs (n, x, y) of positive integers such that m, n are relatively prime and $(x^2 + y^2)^m = (xy)^n$, where n, x, y can be represented by functions of m.

solution page 244

8 th Korean Mathematical Olympiad (Final Round)

1

For any positive integer m, show that there exist integers a, b satisfying

$$|a| \leq m, \quad |b| \leq m, \quad 0 < a + b\sqrt{2} \leq \frac{1 + \sqrt{2}}{m + 2}.$$

solution page 245

Israel Mathematical Olympiad, 1995

7

Solve the system

$$x + \log(x + \sqrt{x^2 + 1}) = y,$$
$$y + \log(y + \sqrt{y^2 + 1}) = z,$$
$$z + \log(z + \sqrt{z^2 + 1}) = x.$$

solution page 246

45th Latvian Mathematical Olympiad, 1994 (11th Grade)

1

Prove for each choice of real non-zero numbers a_1, a_2, a_3, the "stars" can be replaced by "$<$" and "$>$" so that the system

$$\begin{cases} a_1 x + b_1 y + c_1 * 0 & (1) \\ a_2 x + b_2 y + c_2 * 0 & (2) \\ a_3 x + b_3 y + c_3 * 0 & (3) \end{cases}$$

has no solution.

solution page 247

45th Latvian Mathematical Olympiad, 1994 (11th Grade)

2

Solve in natural numbers:

$$x(x + 1) = y^7.$$

solution page 248

45th Latvian Mathematical Olympiad, 1994 (12th Grade)

1

Solve the equation $\cos x \cdot \cos 2x \cdot \cos 3x = 1$.

solution page 249

XI Italian Mathematical Olympiad, 1995

6

Find all pairs of positive integers x, y such that

$$x^2 + 615 = 2^y.$$

solution page 250

Georgian Mathematical Olympiad, 1995 (Final Round)

3

solution page 251

Prove that for any natural number n, the average of all its factors lies between the numbers $\sqrt{n}$ and $\frac{n+1}{2}$.

Bi-National Israel-Hungary Mathematical Competition, 1995

1

solution page 252

Denote the sum of the first n prime numbers by S_n. Prove that there exists a whole square between S_n and S_{n+1}.

31st Spanish Mathematical Olympiad, 1994 (First Round)

2

solution page 253

Show that, if $(x + \sqrt{x^2 + 1})(y + \sqrt{y^2 + 1}) = 1$, then $x + y = 0$.

31st Spanish Mathematical Olympiad, 1994 (First Round)

4

solution page 254

Find the smallest natural number m such that, for all natural numbers $n \geq m$, we have $n = 5a + 11b$, with a, b integers ≥ 0.

31st Spanish Mathematical Olympiad, 1995 (Second Round)

4

solution page 255

Find all the integer solutions of the equation

$$p(x + y) = xy$$

in which p is a prime number.

31st Spanish Mathematical Olympiad, 1995 (Second Round)

5

solution page 256

Show that, if the equations

$$x^3 + mx - n = 0, \quad nx^3 - 2m^2x^2 - 5mnx - 2m^3 - n^2 = 0 \quad (m \neq 0, n \neq 0)$$

have a common root, then the first equation would have two equal roots, and determine in this case the roots of both equations in terms of n.

28th Austrian Mathematical Olympiad, 1997

1

solution page 257

Let a be a fixed whole number.
   Determine all solutions x, y, z in whole numbers to the system of equations

$$
\begin{aligned}
5x + (a+2)y + (a+2)z &= a, \\
(2a+4)x + (a^2+3)y + (2a+2)z &= 3a - 1, \\
(2a+4)x + (2a+2)y + (a^2+3)z &= a + 1.
\end{aligned}
$$

28th Austrian Mathematical Olympiad, 1997

4

Determine all quadruples (a, b, c, d) of real numbers satisfying the equation

$$256a^3b^3c^3d^3 = (a^6 + b^2 + c^2 + d^2)(a^2 + b^6 + c^2 + d^2)$$
$$\times (a^2 + b^2 + c^6 + d^2)(a^2 + b^2 + c^2 + d^6).$$

solution page 258

Iranian Mathematical Olympiad, 1997 (Final Round)

3

Suppose that $w_1, w_2, \ldots, w_k$ are distinct real numbers with a non-zero sum. Prove that there exist integer numbers $n_1, n_2, \ldots, n_k$ such that $\sum_{i=1}^{k} n_i w_i > 0$ and for any non-identity permutation π on $\{1, 2, \ldots, k\}$ we have $\sum_{i=1}^{k} n_i w_{\pi(i)} < 0$.

solution page 259

Ukrainian Mathematical Olympiad, 1998 (9th Grade)

1

Prove the inequality

$$\frac{1 + ab}{1 + a} + \frac{1 + bc}{1 + b} + \frac{1 + ac}{1 + c} \geq 3$$

for positive real numbers a, b, c with $abc = 1$.

solution page 260

Ukrainian Mathematical Olympiad, 1998 (11th Grade)

5

For real numbers $x, y, z \in (0, 1]$, prove the inequality

$$\frac{x}{1 + y + zx} + \frac{y}{1 + z + xy} + \frac{z}{1 + x + yz} \leq \frac{3}{x + y + z}.$$

solution page 261

Ukrainian Mathematical Olympiad, 1998 (11th Grade)

8

Let $x_1, x_2, \ldots, x_n, \ldots$ be the sequence of real numbers such that

$$x_1 = 1, \quad x_{n+1} = \frac{n^2}{x_n} + \frac{x_n}{n^2} + 2, \quad n \geq 1.$$

Prove that

(a) $x_{n+1} \geq x_n$ for all $n \geq 4$;

(b) $[x_n] = n$ for all $n \geq 4$ ($[a]$ denotes the whole part of a).

solution page 262

Vietnamese Mathematical Olympiad, 1998 (Category B, Day 1) 4

Let $x_1, x_2, \ldots, x_n$ $(n \geq 2)$ be real positive numbers satisfying

$$\frac{1}{x_1 + 1998} + \frac{1}{x_2 + 1998} + \cdots + \frac{1}{x_n + 1998} = \frac{1}{1998}.$$

Prove that

$$\frac{\sqrt[n]{x_1 x_2 \cdots x_n}}{n - 1} \geq 1998.$$

solution page 264

15th Balkan Mathematical Olympiad, 1998 2

Let n be an integer, $n \geq 2$, and $0 < a_1 < a_2 < \cdots < a_{2n+1}$ be real numbers. Prove that the following inequality holds:

$$\sqrt[n]{a_1} - \sqrt[n]{a_2} + \sqrt[n]{a_3} - \cdots - \sqrt[n]{a_{2n}} + \sqrt[n]{a_{2n+1}}$$
$$< \sqrt[n]{a_1 - a_2 + a_3 - \cdots - a_{2n} + a_{2n+1}}.$$

solution page 265

15th Balkan Mathematical Olympiad, 1998 4

Prove that the equation $y^2 = x^5 - 4$ has no integer solutions.

solution page 266

Final National Selection Competition for Greek Team, 1998 1

If $x, y, z > 0$, $k > 2$ and $a = x + ky + kz$, $b = kx + y + kz$, $c = kx + ky + z$, show that

$$\frac{x}{a} + \frac{y}{b} + \frac{z}{c} \geq \frac{3}{2k + 1}.$$

solution page 267

47th Czech and Slovak Mathematical Olympiad, 1998 1

Find all solutions in the real domain of the equation

$$x \cdot \lfloor x \cdot \lfloor x \cdot \lfloor x \rfloor \rfloor \rfloor = 88,$$

where $\lfloor a \rfloor$ is the integer part of a real number a; that is, the integer satisfying $\lfloor a \rfloor \leq a < \lfloor a \rfloor + 1$. For instance, $\lfloor 3.7 \rfloor = 3$, $\lfloor -3.7 \rfloor = -4$ and $\lfloor 6 \rfloor = 6$.

solution page 268

47th Czech and Slovak Mathematical Olympiad, 1998

6

Let a, b, c be positive numbers. Show that the triangle with sides a, b, c exists if and only if the system of equations

$$\frac{y}{z} + \frac{z}{y} = \frac{a}{x}, \quad \frac{z}{x} + \frac{x}{z} = \frac{b}{y}, \quad \frac{x}{y} + \frac{y}{x} = \frac{c}{z}$$

has a solution in the real domain.

solution page 270

USA Team Selection Test for IMO, 1999

6

Solve the equation

$$\frac{1}{x^2} + \frac{1}{(4 - \sqrt{3}\,x)^2} = 1.$$

solution page 271

Grosman Memorial Mathematical Olympiad, 1999

2

Find the smallest integer n for which $0 < \sqrt[3]{n} - \lfloor\sqrt[3]{n}\rfloor < 10^{-5}$.

solution page 272

16th Iranian Mathematical Olympiad, 1999 (First Round)

1

Suppose that $a_1 < a_2 < \cdots < a_n$ are real numbers. Prove that:

$$a_1 a_2^4 + a_2 a_3^4 + \cdots + a_{n-1} a_n^4 + a_n a_1^4 \geq a_2 a_1^4 + a_3 a_2^4 + \cdots + a_n a_{n-1}^4 + a_1 a_n^4.$$

solution page 273

16th Iranian Mathematical Olympiad, 1999 (First Round)

5

Suppose that n is a positive integer and $d_1 < d_2 < d_3 < d_4$ are the four smallest positive integers dividing n. Find all integers n satisfying $n = d_1^2 + d_2^2 + d_3^2 + d_4^2$.

solution page 274

16th Iranian Mathematical Olympiad, 1999 (Third Round)

6

Suppose that r_1, ..., r_n are real numbers. Prove that there exists $I \subseteq \{1, 2, \ldots, n\}$ such that I meets $\{i, i + 1, i + 2\}$ in at least one and at most two elements, for $1 \leq i \leq n - 2$ and

$$\left| \sum_{i \in I} r_i \right| \geq \frac{1}{6} \sum_{i=1}^{n} |r_i|.$$

solution page 275

Chinese Mathematical Olympiad, 1999

4

Let m be a given integer. Prove that there exist integers a, b, and k such that both a, b are not divisible by 2, $k \geq 0$, and

$$2m = a^{19} + b^{99} + k \cdot 2^{1999}.$$

solution page 276

Vietnamese Mathematical Olympiad, 1999 (Category A)

1

Solve the system of equations

$$\begin{cases} (1 + 4^{2x-y})5^{1-2x+y} & = 1 + 2^{2x-y+1} \\ y^3 + 4x + 1 + \ln(y^2 + 2x) & = 0 \end{cases}$$

solution page 277

Vietnamese Mathematical Olympiad, 1999 (Category A)

3

Let $\{x_n\}_{n=0}^{\infty}$ and $\{y_n\}_{n=0}^{\infty}$ be two sequences defined recursively as follows:

$$x_0 = 1, \quad x_1 = 4, \quad x_{n+2} = 3x_{n+1} - x_n,$$
$$y_0 = 1, \quad y_1 = 2, \quad y_{n+2} = 3y_{n+1} - y_n,$$

for all $n = 0, 1, 2, \ldots$.

(a) Prove that

$$x_n^2 - 5y_n^2 + 4 = 0$$

for all non-negative integers n.

(b) Suppose that a, b are two positive integers such that $a^2 - 5b^2 + 4 = 0$. Prove that there exists a non-negative integer k such that $x_k = a$ and $y_k = b$.

solution page 278

6th Turkish Mathematical Olympiad, 1999 (Second Round)

2

Prove that

$$(a + 3b)(b + 4c)(c + 2a) \geq 60abc$$

for all real numbers $0 \leq a \leq b \leq c$.

solution page 280

6th Turkish Mathematical Olympiad, 1999 (Second Round)

4

Determine all positive integers x, n satisfying the equation $x^3 + 3367 = 2^n$.

solution page 281

Swiss Mathematical Contest, 1999 (First Day)

4

Find all solutions $(x, y, z) \in \mathbb{R} \times \mathbb{R} \times \mathbb{R}$ of the system

$$\frac{4x^2}{1 + 4x^2} = y, \quad \frac{4y^2}{1 + 4y^2} = z, \quad \frac{4z^2}{1 + 4z^2} = x.$$

solution page 282

50th Polish Mathematical Olympiad, 1999

Find all integers $n \geq 2$ for which the system of equations

$$
\begin{cases}
x_1^2 + x_2^2 + 50 &= 16x_1 + 12x_2 \\
x_2^2 + x_3^2 + 50 &= 16x_2 + 12x_3 \\
x_3^2 + x_4^2 + 50 &= 16x_3 + 12x_4 \\
\cdots\cdots\cdots & \cdots \quad \cdots\cdots \\
x_{n-1}^2 + x_n^2 + 50 &= 16x_{n-1} + 12x_n \\
x_n^2 + x_1^2 + 50 &= 16x_n + 12x_1
\end{cases}
$$

has a solution in integers x_1, x_2, x_3, $\ldots$, x_n.

solution page 283

St. Petersburg Mathematical Contest

Does there exist a positive integer n such that

$$27^n + 84^n + 110^n + 133^n = 144^n ?$$

solution page 284

St. Petersburg Mathematical Contest

Let a, b and c be real numbers with sum 0. Prove that

$$\frac{a^7 + b^7 + c^7}{7} = \left(\frac{a^5 + b^5 + c^5}{5}\right)\left(\frac{a^2 + b^2 + c^2}{2}\right).$$

solution page 285

St. Petersburg Mathematical Contest

Prove that

$$\sum_{k=0}^{n} \binom{n}{k}(a+k)^{k-1}(b+n-k)^{n-k-1} = (a+b+n)^{n-1}\left(\frac{1}{a} + \frac{1}{b}\right).$$

solution page 286

Ukranian Mathematical Olympiad, 1999 (10th Grade)

For real numbers $x_1, x_2, \ldots, x_6 \in [0,1]$ prove the inequality

$$
\frac{x_1^3}{x_2^5 + x_3^5 + x_4^5 + x_5^5 + x_6^5 + 5} + \frac{x_2^3}{x_1^5 + x_3^5 + x_4^5 + x_5^5 + x_6^5 + 5} + \cdots
$$
$$
+ \frac{x_6^3}{x_1^5 + x_2^5 + x_3^5 + x_4^5 + x_5^5 + 5} \leq \frac{3}{5}.
$$

solution page 288

XLIII Mathematical Olympiad of Moldova, 1999 (10th Form)

Prove that for all strictly positive numbers a, b, and c the inequality

$$(a + b + x)^{-1} + (b + c + x)^{-1} + (c + a + x)^{-1} \leq x^{-1},$$

holds, where $x = \sqrt[3]{abc}$.

solution page 289

XLIII Mathematical Olympiad of Moldova, 1999 (11th Form)

Find all the integer values of m, for which the equation

$$\left\lfloor \frac{m^2 x - 13}{1999} \right\rfloor = \frac{x - 12}{2000}$$

has 1999 distinct real solutions ($\lfloor \cdot \rfloor$ denotes the integral part function).

solution page 290

Italian Team Selection Test, 1999

Prove that for each prime number p the equation

$$2^p + 3^p = a^n$$

has no solutions (a, n), with a and n integers > 1.

solution page 291

XV Gara Nazionale di Matematica, 1999

(a) Determine all pairs (x, k) of positive integers which satisfy the equation

$$3^k - 1 = x^3.$$

(b) Prove that if n is an integer greater than 1 and different from 3 there are no pairs (x, k) of positive integers satisfying the equation

$$3^k - 1 = x^n.$$

solution page 292

Chinese Mathematical Olympiad, 2000

Find all positive integers n for which there are k integers $n_1, n_2, \ldots, n_k$, each greater than 3, such that

$$n = n_1 n_2 \cdots n_k = \sqrt[2^k]{2^{(n_1 - 1)(n_2 - 1)\cdots(n_k - 1)}} - 1.$$

solution page 293

Prove that there exist ten different real numbers $a_1, a_2, \ldots, a_{10}$ such that the equation

$$(x - a_1)(x - a_2) \cdots (x - a_{10}) = (x + a_1)(x + a_2) \cdots (x + a_{10})$$

has exactly 5 different real roots.

solution page 294

Let $a_1, a_2, \ldots, a_{2000}$ be real numbers such that
$$a_1^3 + a_2^3 + \cdots + a_n^3 = (a_1 + a_2 + \cdots + a_n)^2$$
for all n, $1 \leq n \leq 2000$. Prove that every element of the sequence is an integer.

solution page 295

Prove that
$$\frac{1}{\sqrt{1 + x^2}} + \frac{1}{\sqrt{1 + y^2}} < x, y \leq 1.$$

solution page 297

Prove that for any prime p, there exist integers x, y, z, and w such that $x^2 + y^2 + z^2 - wp = 0$ and $0 < w < p$.

solution page 298

The real numbers a, b, c, x, y, and z are such that $a > b > c > 0$ and $x > y > z > 0$. Prove that

$$\frac{a^2 x^2}{(by + cz)(bz + cy)} + \frac{b^2 y^2}{(cz + ax)(cx + az)} + \frac{c^2 z^2}{(ax + by)(ay + bx)} \geq \frac{3}{4}.$$

solution page 299

Pete and Bill play the following game. At the beginning, Pete chooses a number a, then Bill chooses a number b, and then Pete chooses a number c. Can Pete choose his numbers in such a way that the three equations $x^3 + ax^2 + bx + c = 0$, $x^3 + bx^2 + cx + a = 0$, and $x^3 + cx^2 + ax + b = 0$ have a common

(a) real root?

(b) negative root?

solution page 301

Belarusian Mathematical Olympiad, 2000

2

How many pairs (n, q) satisfy $\{q^2\} = \left\{ \dfrac{n!}{2000} \right\}$, where n is a positive integer and q is a non-integer rational number such that $0 < q < 2000$? $\{r\}$ means the "fractional part" of r.

solution page 302

Belarusian Mathematical Olympiad, 2000

7

(a) Find all positive integers n such that $(a^a)^n = b^b$ has at least one solution in integers a and b, both exceeding 1.

(b) Find all positive integers a and b such that $(a^a)^5 = b^b$.

solution page 304

Taiwanese Mathematical Olympiad, 2000

1

Find all pairs (x, y) of positive integers such that $y^{x^2} = x^{y+2}$.

solution page 305

Russian Mathematical Olympiad, 2000

5

Prove that
$$\frac{1}{\sqrt{1 + x^2}} + \frac{1}{\sqrt{1 + y^2}} \leq \frac{2}{\sqrt{1 + xy}} \quad \text{for } 0 < x, y \leq 1.$$

solution page 306

32nd Austrian Mathematical Olympiad

2

Determine all triplets of positive real numbers x, y, and z solving the system of equations
$$x + y + z = 6,$$
$$\frac{1}{x} + \frac{1}{y} + \frac{1}{z} = 2 - \frac{4}{xyz}.$$

solution page 307

32nd Austrian Mathematical Olympiad

5

Determine all whole numbers m for which all solutions of the equation $3x^3 - 3x^2 + m = 0$ are rational numbers.

solution page 308

Finish High School Mathematical Competition, 2000

3

Determine all positive integers n such that $n! > \sqrt{n^n}$.

solution page 309

Ukrainian Mathematical Olympiad, 2001 (Grade 10)

3

Let $a_1, a_2, \ldots, a_n$ be real numbers such that

$$a_1 + a_2 + \cdots + a_n \geq n^2 \quad \text{and} \quad a_1^2 + a_2^2 + \cdots + a_n^2 \leq n^3 + 1.$$

Prove that $n - 1 \leq a_k \leq n + 1$ for all k.

solution page 510

Ukrainian Mathematical Olympiad, 2001 (Grade 11)

8

Let a, b, c and α, β, γ be positive real numbers such that $\alpha + \beta + \gamma = 1$. Prove the inequality

$$\alpha a + \beta b + \gamma c + 2\sqrt{(\alpha\beta + \beta\gamma + \gamma\alpha)(ab + bc + ca)} \leq a + b + c.$$

solution page 511

XVII National Mathematical Contest of Italy

3

Given the equation $x^{2001} = y^x$,

(a) find all solution pairs (x, y) consisting of positive integers with x prime;

(b) find all solution pairs (x, y) consisting of positive integers.

(Recall that $2001 = 3 \cdot 23 \cdot 29$.)

solution page 512

52nd Polish Mathematical Olympiad

1

Show that the inequality

$$\sum_{i=1}^{n} ix_i \leq \binom{n}{2} + \sum_{i=1}^{n} x_i^i$$

holds for every integer $n \geq 2$ and all real numbers $x_1, x_2, \ldots, x_n \geq 0$.

solution page 513

Hungary-Israel Mathematical Competition, 2001 (Individual)

1

Find positive integers x, y, z such that $x > z > 1999 \cdot 2000 \cdot 2001 > y$ and $2000x^2 + y^2 = 2001z^2$.

solution page 514

49th Mathematical Olympiad of Lithuania, 2000

8

The equation $x^2 + y^2 + z^2 + u^2 = xyzu + 6$ is given. Find:

(a) at least one solution in positive integers;

(b) at least 33 such solutions;

(c) at least 100 such solutions.

solution page 515

Hungarian Mathematical Olympiad, 2000 (First Round)

1

Let x, y, and z denote positive real numbers, each less than 4. Prove that at least one of the numbers $\dfrac{1}{x} + \dfrac{1}{4-y}$, $\dfrac{1}{y} + \dfrac{1}{4-z}$, and $\dfrac{1}{z} + \dfrac{1}{4-x}$ is greater than or equal to 1.

solution page 516

Hungarian Mathematical Olympiad, 2000 (First Round)

2

Find the integer solutions of $5x^2 - 14y^2 = 11z^2$.

solution page 517

11th Japanese Mathematical Olympiad, (Second Round)

3

Three real numbers a, b, $c \geq 0$ satisfy

$$a^2 \leq b^2 + c^2, \qquad b^2 \leq c^2 + a^2, \qquad c^2 \leq a^2 + b^2.$$

Prove the inequality

$$(a + b + c)(a^2 + b^2 + c^2)(a^3 + b^3 + c^3) \geq 4(a^6 + b^6 + c^6).$$

When does equality hold?

solution page 518

XXI Albanian Mathematical Olympiad, 2000 (12th Form)

1

(a) Prove the inequality

$$\frac{(1 + x_1)(1 + x_2) \cdots (1 + x_n)}{1 + x_1 x_2 \cdots x_n} \leq 2^{n-1}, \quad \forall x_1, x_2, \ldots, x_n \in [1, +\infty).$$

(b) When does the equality hold?

solution page 519

XXI Albanian Mathematical Olympiad, 2000 (12th Form)

3

Prove that, if $0 < a < b < \dfrac{\pi}{2}$, then

(a) $\dfrac{a}{b} < \dfrac{\sin a}{\sin b}$;

(b) $\dfrac{\sin a}{\sin b} < \dfrac{\pi}{2} \dfrac{a}{b}$.

solution page 520

Finland Mathematical Olympiad, 2000 (1st Round)

3

Determine all positive integers m and n such that

$$m^2 - n^2 = 270.$$

solution page 521

Finland Mathematical Olympiad, 2000 (Final Round)

3

The positive integers a, b, and c satisfy $\frac{1}{a} + \frac{1}{b} + \frac{1}{c} < 1$. Prove that

$$\frac{1}{a} + \frac{1}{b} + \frac{1}{c} \leq \frac{41}{42}.$$

solution page 322

8th Macedonian Mathematical Olympiad

1

Prove that, if $m \cdot s = 2000^{2001}$ where m, $s \in \mathbb{Z}$, then the equation $mx^2 - sy^2 = 3$ has no solution in $\mathbb{Z}$.

solution page 323

13th Irish Mathematical Olympiad

6

Let $x \geq 0$, $y \geq 0$ be real numbers with $x + y = 2$. Prove that

$$x^2 y^2 (x^2 + y^2) \leq 2.$$

solution page 324

13th Irish Mathematical Olympiad

8

For each positive integer n, determine, with proof, all positive integers m such that there exist positive integers $x_1 < x_2 < \cdots < x_n$ which satisfy $\frac{1}{x_1} + \frac{2}{x_2} + \frac{3}{x_3} + \cdots + \frac{n}{x_n} = m$.

solution page 325

Third Hong Kong Mathematical Olympiad

2

Let $a_1 = 1$, $a_{n+1} = \frac{a_n}{n} + \frac{n}{a_n}$ for $n = 1, 2, 3, \ldots$. Find the greatest integer less than or equal to a_{2000}. Be sure to prove your claim.

solution page 326

Icelandic Mathematical Contest, 2000

1

Let x and y be positive real numbers such that $xy = 1$. Prove that

$$\frac{x}{y} + \frac{y}{x} \geq 2.$$

solution page 327

Greek Team Selection Test for IMO 2002

2

Let x, y, a be real numbers such that

$$x + y = x^3 + y^3 = x^5 + y^5 = a.$$

Determine all the possible values of a.

solution page 328

Prove that the following inequality holds for every triple (a, b, c) of non-negative real numbers with $a^2 + b^2 + c^2 = 1$:

$$\frac{a}{b^2 + 1} + \frac{b}{c^2 + 1} + \frac{c}{a^2 + 1} \geq \frac{3}{4} \left(a\sqrt{a} + b\sqrt{b} + c\sqrt{c} \right)^2.$$

solution page 329

When does equality hold?

Find all positive integers z for which the equation

$$x(x + z) = y^2$$

has no solutions x, y that are positive integers.

solution page 330

Let x and y be any two real numbers. Prove that

$$3(x + y + 1)^2 + 1 \geq 3xy.$$

Under what conditions does equality hold?

solution page 331

Determine all triples of positive integers a, b, c such that $a^2 + 1$ and $b^2 + 1$ are prime numbers satisfying $(a^2 + 1)(b^2 + 1) = c^2 + 1$.

solution page 332

Prove that, for every integer $n \geq 3$ and every sequence of positive numbers $x_1, x_2, \ldots, x_n$, at least one of the following two inequalities is satisfied:

$$\sum_{i=1}^{n} \frac{x_i}{x_{i+1} + x_{i+2}} \geq \frac{n}{2}, \qquad \sum_{i=1}^{n} \frac{x_i}{x_{i-1} + x_{i-2}} \geq \frac{n}{2}.$$

(Note: Here $x_{n+1} = x_1$, $x_{n+2} = x_2$, $x_0 = x_n$, and $x_{-1} = x_{n-1}$.)

solution page 334

Singapore Mathematical Olympiad, 2002 (Open Section, Part B)

2

Let $a_1, a_2, \ldots, a_n$ and $b_1, b_2, \ldots, b_n$ be real numbers between 1001 and 2002 inclusive. Suppose $\sum_{i=1}^{n} a_i^2 = \sum_{i=1}^{n} b_i^2$. Prove that

$$\sum_{i=1}^{n} \frac{a_i^3}{b_i} \leq \frac{17}{10} \sum_{i=1}^{n} a_i^2 .$$

Determine when equality holds.

solution page 335

XVIII Italian Mathematical Olympiad, 2002

4

Find all values of n for which all solutions of the equation $x^3 - 3x + n = 0$ are integers.

solution page 336

British Mathematical Olympiad, 2001 (Round 1)

1

Find all positive integers m, n, where n is odd, that satisfy

$$\frac{1}{m} + \frac{4}{n} = \frac{1}{12} .$$

solution page 337

British Mathematical Olympiad, 2001 (Round 1)

3

Find all positive real solutions to the equation

$$x + \left\lfloor \frac{x}{6} \right\rfloor = \left\lfloor \frac{x}{2} \right\rfloor + \left\lfloor \frac{2x}{3} \right\rfloor ,$$

where $\lfloor t \rfloor$ denotes the largest integer less than or equal to the real number t.

solution page 338

Yugoslav Mathematical Olympiad, 2002

1

Let a, b, and c be positive numbers, and let n and k be positive integers. Prove the inequality:

$$\frac{a^{n+k}}{b^n} + \frac{b^{n+k}}{c^n} + \frac{c^{n+k}}{a^n} \geq a^k + b^k + c^k .$$

solution page 339

Yugoslav Qualification for IMO, 2002 (Second Round)

1

What is the maximal value of the expression $a + b + c + abc$, if a, b, c are non-negative numbers such that $a^2 + b^2 + c^2 + abc \leq 4$?

solution page 340

Let x, y, and z be real numbers such that
$$x + y + z = 3 \quad \text{and} \quad xy + yz + xz = a$$
(a is a real parameter). Determine the value of the parameter a for which the difference between the maximum and minimum possible values of x equals 8.

solution page 341

Let a, b, and c be real numbers such that $a^2 + b^2 + c^2 = 1$. Prove the inequality
$$\frac{a^2}{1 + 2bc} + \frac{b^2}{1 + 2ca} + \frac{c^2}{1 + 2ab} \geq \frac{3}{5}.$$

solution page 342

Let p and q be different prime numbers. Solve the following system of equations in the set of integers:
$$\frac{z+p}{x} + \frac{z-p}{y} = q,$$
$$\frac{z+p}{y} - \frac{z-p}{x} = q.$$

solution page 343

Find all positive integers n such that the equation $x^3 + y^3 + z^3 = nx^2y^2z^2$ has positive integer solutions. Be sure to give a proof.

solution page 345

Find all triples of positive integers (p, q, n), with p and q prime, such that
$$p(p + 3) + q(q + 3) = n(n + 3).$$

solution page 346

Let $0 < a, b, c < 1$. Prove that
$$\frac{a}{1-a} + \frac{b}{1-b} + \frac{c}{1-c} \geq \frac{3\sqrt[3]{abc}}{1 - \sqrt[3]{abc}}.$$
Determine the case of equality.

solution page 347

15th Irish Mathematical Olympiad, (First Paper)

7

Suppose n is a product of four distinct primes a, b, c, d such that

(a) $a + c = d$;

(b) $a(a + b + c + d) = c(d - b)$;

(c) $1 + bc + d = bd$.

Determine n.

solution page 548

38th Mongolian Mathematical Olympiad, (Final Round)

5

Let a_0, a_1, ... be an infinite sequence of positive real numbers. Show that $1 + a_n > \sqrt[n]{2}\, a_{n-1}$ for infinitely many positive integers n.

solution page 549

XXIX Russian Mathematical Olympiad, Final Round 11th Form

1

Let α, β, γ, and τ be positive numbers such that, for all x,

$$\sin \alpha x + \sin \beta x = \sin \gamma x + \sin \tau x.$$

Prove that $\alpha = \gamma$ or $\alpha = \tau$.

solution page 550

Japan Mathematical Olympiad, 2003

3

Find the greatest real number k such that, for any positive a, b, c with $a^2 > bc$,

$$(a^2 - bc)^2 > k(b^2 - ca)(c^2 - ab).$$

solution page 551

Kűrschák Mathematical Competition, 2002

5

The Fibonacci sequence is defined by the following recursion: $f_1 = f_2 = 1$ and $f_n = f_{n-1} + f_{n-2}$ for $n > 2$. Suppose that the positive integers a and b satisfy:

$$\min\left\{\frac{f_n}{f_{n-1}}, \frac{f_{n+1}}{f_n}\right\} \le \frac{a}{b} \le \max\left\{\frac{f_n}{f_{n-1}}, \frac{f_{n+1}}{f_n}\right\}.$$

Prove that $b \ge f_{n+1}$.

solution page 552

British Mathematical Competition, 2003 (Round I)

3

Let x, y, z be positive real numbers such that $x^2 + y^2 + z^2 = 1$. Prove that

$$x^2yz + xy^2z + xyz^2 \le \tfrac{1}{3}.$$

solution page 554

Find all solutions in positive integers a, b, c to the equation

$$a!b! = a! + b! + c!$$

solution page 355

For each integer $n > 1$, let $p(n)$ denote the largest prime factor of n. Determine all triples x, y, z of distinct positive integers satisfying

(i) x, y, z are in arithmetic progression, and

(ii) $p(xyz) \leq 3$.

solution page 356

Let $f : \mathbb{N} \to \mathbb{N}$ be a permutation of the set $\mathbb{N}$ of all positive integers.

(a) Show that there is an arithmetic progression a, $a + d$, $a + 2d$, where $d > 0$, such that $f(a) < f(a + d) < f(a + 2d)$.

(b) Must there be an arithmetic progression a, $a+d$, $\ldots$, $a+2003d$, where $d > 0$, such that $f(a) < f(a + d) < \cdots < f(a + 2003d)$?

[A permutation of $\mathbb{N}$ is a one-to-one function whose image is the whole of $\mathbb{N}$; that is, a function from $\mathbb{N}$ to $\mathbb{N}$ such that for all $m \in \mathbb{N}$ there is a unique $n \in \mathbb{N}$ such that $f(n) = m$.]

solution page 357

Find all real k such that the following system of equations has a unique solution:

$$x^2 + y^2 = 2k^2,$$
$$kx - y = 2k.$$

solution page 358

Does there exist a number $q \in \mathbb{N}$ and a prime number $p \in \mathbb{N}$ such that

$$3^p + 7^p = 2 \cdot 5^q?$$

solution page 359

German Mathematical Olympiad, 2003

1

Determine all pairs (x, y) of real numbers x, y which satisfy

$$x^3 + y^3 = 7,$$
$$xy(x + y) = -2.$$

solution page 360

Iranian Mathematical Olympiad, 2002 (Second Round)

6

Let a, b, and c be positive real numbers such that $a^2 + b^2 + c^2 + abc = 4$. Prove that $a + b + c \leq 3$.

solution page 361

algebra
problems and <u>solutions</u>
from
Mathematical Olympiads

Find the number of those subsets of $\{1, 2, \ldots, 2n\}$ in which the equation $x + y = 2n + 1$ has no solutions.

Solution

That the equation $x + y = 2n + 1$ has no solution means that x and $2n + 1 - x$ are not both chosen from $x = 1, 2, \ldots, n$. Hence we have three possibilities: to choose one, the other, or neither. The answer is 3^n subsets (counting the empty set).

Solve the following equation in positive integers:

$$x^3 - y^3 = xy + 61.$$

Solution

Manipulating the equation we obtain

$$(x - y)(x^2 + xy + yz) - xy = 61$$

and

$$(x - y)((x - y)^2 + 3yx) - xy = 61.$$

Set $x - y = a$. Since the right-hand side is positive, $a = x - y > 0$. Set $b = xy$. Note that $b > 0$.

Rewriting the equation in terms of a and b yields

$$a^3 + b(3a - 1) = 61,$$

where $a > 0$ and $b > 0$.

Thus $a^3 < 61$, so $a = 1, 2, 3$.

Also $b = \frac{61 - a^3}{3a - 1} \in \mathbb{Z}$, so trying

$$
\begin{aligned}
a &= 1 \quad \text{gives} \quad b = 30, \\
a &= 2 \quad \text{gives} \quad b = 53/5, \\
a &= 3 \quad \text{gives} \quad b = 17/4.
\end{aligned}
$$

Thus $a = 1$, $x - y = 1$ and $xy = 30$, giving $(y + 1)y = 30$ with $y = 5$ (rejecting $y = -6$) and $x = 6$.

The unique solution is $x = 6, y = 5$.

Sequences $a_1, \ldots, a_n, \ldots$ and $b_1, \ldots, b_n, \ldots$ are such that $a_1 > 0$, $b_1 > 0$, and

$$a_{n+1} = a_n + \frac{1}{b_n}, \quad b_{n+1} = b_n + \frac{1}{a_n}, \quad n \in \mathbb{N}.$$

Prove that

$$a_{25} + b_{25} > 10\sqrt{2}.$$

Solution

It follows that $a_2 + b_2 \geq 4$ and

$$a_{n+1} + b_{n+1} = (a_n + b_n)(1 + 1/a_n b_n) \geq (a_n + b_n) + 4/(a_n + b_n).$$

Let $x_{n+1} = x_n + 4/x_n$ with $x_2 = 4$. Then since $x + 4/x$ is increasing for $x \geq 2$, we have $a_{n+1} + b_{n+1} \geq x_{n+1}$. Since $(x_{n+1})^2 = (x_n)^2 + (4/x_n)^2 + 8$,

$$(x_{n+1})^2 \leq (x_n)^2 + 8.$$

Summing the latter set of inequalities for $n = 2$ to $n - 1$, we obtain

$$(x_n)^2 \leq x_2^2 + 8(n - 2) = 8n.$$

It then follows that

$$(x_{n+1})^2 \geq (x_n)^2 + 8 + 1/2n,$$

and summing this inequality for $n = 2$ to 24, we obtain

$$(x_{25})^2 \geq (x_2)^2 + 8(23) + (1/4 + 1/6 + \cdots + 1/48).$$

Hence, $(x_{25})^2 > 16 + 184$ or $10\sqrt{2} < x_{25} \leq a_{25} + b_{25}$.

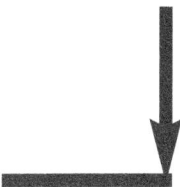

Next we give the generalization by Seiffert.

More generally: Let $u > 0$, $v > 0$, and $w > 0$. If the sequences $\{a_n\}_{n\geq 1}$ and $\{b_n\}_{n\geq 1}$ satisfy $a_1 > 0$, $b_1 > 0$, and

$$a_{n+1} = ua_n + \frac{v}{b_n}, \quad b_{n+1} = \frac{b_n}{u} + \frac{w}{a_n}, \quad n \in N,$$

then

$$a_n b_n > (n-1)\left(\frac{v}{u} + uw\right) + 2\sqrt{vw}, \quad n \geq 3, \tag{1}$$

and

$$a_n + b_n > 2\sqrt{(n-1)\left(\frac{v}{u} + uw\right) + 2\sqrt{vw}}, \quad n \geq 3. \tag{2}$$

First, we note that $a_n > 0$ and $b_n > 0$, $n \in \mathbb{N}$. We have

$$a_{k+1}b_{k+1} = \left(ua_k + \frac{v}{b_k}\right)\left(\frac{b_k}{u} + \frac{w}{a_k}\right)$$

$$= \frac{v}{u} + uw + a_k b_k + \frac{vw}{a_k b_k}, \quad k \in \mathbb{N}.$$

Summing as k ranges from 1 to $n-1$, where $n \geq 3$, gives

$$a_n b_n = (n-1)\left(\frac{v}{u} + uw\right) + a_1 b_1 + \sum_{k=1}^{n-1} \frac{vw}{a_k b_k}$$

$$> (n-1)\left(\frac{v}{u} + uw\right) + a_1 b_1 + \frac{vw}{a_1 b_1}$$

$$\geq (n-1)\left(\frac{v}{u} + uw\right) + 2\sqrt{vw},$$

where we have used the AM-GM-Inequality. This proves (1). Then (2) follows from (1) and $a_n + b_n \geq 2\sqrt{a_n b_n}$.

In the particular case $u = v = w = 1$, (1) and (2) give

$$a_n b_n > 2n \quad \text{and} \quad a_n + b_n > 2\sqrt{2n}, \quad n \geq 3.$$

With $n = 25$, we then have $a_{25}b_{25} > 50$ and $a_{25} + b_{25} > 10\sqrt{2}$.

Remark. If $u = 1$ and $v = w$, then $a_1 b_n = b_1 a_n$ for all $n \in \mathbb{N}$, as is easily verified by induction on n.

Two pupils are playing the following game. In the system

$$\begin{cases} *x + *y + *z = 0, \\ *x + *y + *z = 0, \\ *x + *y + *z = 0, \end{cases}$$

they alternately replace the asterisks by any numbers. The first player wins if the final system has a non-zero solution. Can the first player always win?

Solution

Yes. Denote the system as follows:

$$a_1x + a_2y + a_3z = 0, \tag{1}$$
$$b_1x + b_2y + b_3z = 0, \tag{2}$$
$$c_1x + c_2y + c_3z = 0. \tag{3}$$

The first player chooses any number for b_2.

Then form the pairs (a_1, c_1), (a_2, c_2), (a_3, c_3), (b_1, b_3).

Each time the second player chooses a number from one pair, then the first player gives the same number to the other member of the pair. Thus at the end $a_1 = c_1$, $a_2 = c_2$, $a_3 = c_3$, $b_1 = b_3$. So (1) and (3) are equivalent.

And, since the system is homogeneous, it is consistent and must have infinitely many solutions, in particular a non-$(0, 0, 0)$ solution, and the first player wins.

Solve the system of equations:

$$\begin{cases} \sqrt{3x}\left(1 + \frac{1}{x+y}\right) = 2\,, \\ \sqrt{7y}\left(1 - \frac{1}{x+y}\right) = 4\sqrt{2}\,. \end{cases}$$

Solution

Note $x \neq 0$, $y \neq 0$ and $x, y > 0$. We have

$$1 + \frac{1}{x+y} = \frac{2}{\sqrt{3x}}\,, \qquad 1 - \frac{1}{x+y} = \frac{4\sqrt{2}}{\sqrt{7y}}\,.$$

Adding and subtracting, we get the equations

$$2 = \frac{2}{\sqrt{3x}} + \frac{4\sqrt{2}}{\sqrt{7y}}\,, \qquad \frac{2}{x+y} = \frac{2}{\sqrt{3x}} - \frac{4\sqrt{2}}{\sqrt{7y}}\,.$$

Dividing by 2 gives

$$1 = \frac{1}{\sqrt{3x}} + \frac{2\sqrt{2}}{\sqrt{7y}}\,, \qquad \frac{1}{x+y} = \frac{1}{\sqrt{3x}} - \frac{2\sqrt{2}}{\sqrt{7y}}\,.$$

Multiplying, we get

$$\frac{1}{x+y} = \frac{1}{3x} - \frac{8}{7y}\,.$$

Thus,

$$7y^2 - 38xy - 24x^2 = 0\,, \qquad (y - 6x)(7y + 4x) = 0\,,$$

giving $y = 6x$ (since $x, y > 0$).

Substituting, we get $1 = \frac{1}{\sqrt{3x}} + \frac{2\sqrt{2}}{\sqrt{6 \cdot 7x}}$, so that

$$\sqrt{3x} = 1 + \frac{2}{\sqrt{7}} = \frac{2 + \sqrt{7}}{\sqrt{7}}\,,$$

$$3x = \frac{4 + 7 + 4\sqrt{7}}{7}\,, \qquad x = \frac{(2 + \sqrt{7})^2}{21}\,, \qquad y = \frac{2(2 + \sqrt{7})^2}{7}\,.$$

We are given four non-negative real numbers a, b, c, d satisfying the condition:

$$2(ab + ac + ad + bc + bd + cd) + abc + abd + acd + bcd = 16 .$$

Prove that:

$$a + b + c + d \geq \frac{2}{3}(ab + ac + ad + bc + bd + cd) .$$

When does equality occur?

Solution

I will prove the following results: Let n be a positive integer, $n \geq 3$. Let $x_1, x_2, \ldots, x_n$ be non-negative real numbers such that:

$$(n-2) \sum_{i<j} x_i x_j + \sum_{i<j<k} x_i x_j x_k = \frac{2}{3} n(n-1)(n-2) .$$

Then

$$\sum_{i=1}^{n} x_i \geq \frac{2}{n-1} \sum_{i<j} x_i x_j . \tag{1}$$

And equality occurs if and only if:

(i) $n = 3$ and (x_1, x_2, x_3) is a permutation of $(0, 2, 2)$, or

(ii) $n \geq 3$ and $x_1 = x_2 = \cdots = x_n = 1$.

(i) First, we prove the statement for $n = 3$. Let x, y, z be non-negative real numbers such that

$$xy + yz + zx + xyz = 4 . \tag{2}$$

With no loss of generality we can suppose that $x \leq y \leq z$. From (2), we deduce that $3x^2 + x^3 \leq 4$, and so, $x \leq 1$. In the same way $3z^2 + z^3 \geq 4$, and so, $z \geq 1$. Moreover

$$y = \frac{4 - xz}{x + z + xz} .$$

Thus, (1) is equivalent to:

$$\begin{aligned} f(x, z) &= x(x + z + xz) + 4 - xz + z(x + z + xz) - x(4 - xz) \\ &\quad - z(4 - xz) - zx(x + z + xz) \\ &\geq 0 . \end{aligned}$$

But it is easy to see that

$$f(x, z) = (x + z - 2)^2 + xz(z - 1)(1 - x) .$$

The result, for $n = 3$, follows easily.

(ii) Secondly, we prove (1) for $n \geq 3$, by induction on $n \geq 3$. From the first step, we know that (1) is true for $n = 3$. Let $n \geq 3$ be a fixed integer. Suppose that (1) holds for n.

Let $x_1, \ldots, x_{n+1}$ be non-negative real numbers such that

$$(n - 1)\sum_{i<j} x_i x_j + \sum_{1<j<k} x_i x_j x_k = \frac{2}{3}(n + 1)n(n - 1).$$

Denote by

$$S_1 = \sum_{i=1}^{n+1} x_i, \qquad S_2 = \sum_{i<j} x_i x_j, \qquad S_3 = \sum_{i<j<k} x_i x_j x_k.$$

Then we know that

$$(n - 1)S_2 + S_3 = \frac{2}{3}(n + 1)n(n - 1), \tag{3}$$

and we want to prove that

$$S_1 \geq \frac{2}{n}S_2. \tag{4}$$

Let

$$P(x) = \prod_{i=1}^{n+1}(x - x_i) = x^{n+1} - S_1 x^n + S_2 x^{n-1} - S_3 x^{n-2} + \cdots.$$

Then

$$P'(x) = (n + 1)\left(x^n - \frac{nS_1}{n + 1}x^{n-1} + \frac{(n - 1)S_2}{n + 1}x^{n-2} \right.$$
$$\left. - \frac{(n - 2)S_3}{n + 1}x^{n-3} + \cdots \right).$$

Since all roots of P are non-negative real numbers, it follows from Rolle's Theorem that all the roots of P' are non-negative real numbers $\alpha_1, \alpha_2, \ldots, \alpha_n$. Thus, $P'(x) = (n + 1)\prod_{i=1}^{n}(x - \alpha_i)$.

Denote by $S_1' = \sum_{i=1}^{n}\alpha_i$, $S_2' = \sum_{i<j}\alpha_i\alpha_j$, $S_3' = \sum_{i<j<k}\alpha_i\alpha_j\alpha_k$. Then $P'(x) = (n + 1)(x^n - S_1'x^{n-1} + S_2'x^{n-2} - S_3'x^{n-3} + \cdots)$. Identifying the coefficients, we deduce that

$$S_1' = \frac{n}{n+1}S_1, \qquad S_2' = \frac{n-1}{n+1}S_2, \qquad S_3' = \frac{n-2}{n+1}S_3.$$

From (3) we have $(n - 2)S_2' + S_3' = \frac{2}{3}n(n - 1)(n - 2)$. Then, by the induction hypothesis (for $\alpha_1, \ldots, \alpha_n$), we have

$$S_1' \geq \frac{2}{n - 1}S_2'.$$

Thus, $\frac{n}{n+1}S_1 \geq \frac{2}{n-1} \cdot \frac{n-1}{n+1}S_2$, which is equivalent to (4). Then (1) holds for $n + 1$, and the induction is complete.

(iii) Thirdly, the equality case in (1). Suppose $x_1 \leq x_2 \leq \cdots \leq x_n$. From the first step, we know that for $n = 3$, equality occurs if and only if $x_1 = 0$, $x_2 = x_3 = 2$, or $x_1 = x_2 = x_3 = 1$.

Suppose that equality occurs in (1) for $n = 4$. Then, with the notation that is used above, $S_1' = S_2'$. Thus, $\alpha_1 = 0$, and $\alpha_2 = \alpha_3 = 2$; or $\alpha_1 = \alpha_2 = \alpha_3 = 1$.

Case 1. $\alpha_1 = 0$ and $\alpha_2 = \alpha_3 = 2$.

Since 0 is a root of P' and is between two non-negative roots of P, then 0 is also a root of P. It follows that 0 is a root with order at least 2 of P. Thus, $x_1 = x_2 = 0$.

Now, (3) gives $x_3 x_4 = 8$, and $S_1 = \frac{2}{3} S_2$ gives $x_3 + x_4 = \frac{16}{3}$, but the discriminant of the resulting quadratic is negative. This is a contradiction.

Case 2. $\alpha_1 = \alpha_2 = \alpha_3 = 1$.

From Rolle's Theorem, we deduce that P has not more than two distinct real roots: $P(x) = (x - a)^\alpha (x - b)^{4-\alpha}$ with $a \leq b$ and $\alpha \in \{1, 2, 3\}$. Then

$$
\begin{aligned}
P'(x) &= (x - a)^{\alpha - 1}(x - b)^{3-\alpha}(4x - (4a - a\alpha + b\alpha)) \\
&= 4(x - 1)^3.
\end{aligned}
$$

If $\alpha \geq 2$, then $a = 1$, and $4a - a\alpha + b\alpha = 4$. Thus, $a = b = 1$.

If $\alpha = 1$, then $b = 1$, and $4a - a\alpha + b\alpha = 4$. Thus, $a = b = 1$.

Conversely, if $x_1 = x_2 = x_3 = x_4 = 1$, then (3) is satisfied and equality occurs in (1). Then, for $n = 4$, equality occurs if and only if $x_1 = x_2 = x_3 = x_4 = 1$.

(iv) For $n \geq 4$, it is an easy induction on $n \geq 4$, using the same reasoning as above. From the second step, if equality occurs in (1) for the value $n + 1$, then it occurs for the value n.

It follows that equality occurs for $n \geq 4$ if and only if $x_1 = \cdots = x_n = 1$.

Now we give the solutions for Category B.

Let x, y, z be three non-negative real numbers satisfying the condition: $xy + yz + zx + xyz = 4$.

Prove that: $x + y + z \geq xy + yz + zx$. When does equality occur?

Solution

More generally, we show that if

$$xy + yz + zx + xyz = 3a + a^3 \quad (a > 0), \tag{1}$$

then

$$x + y + z \geq xy + yz + zx, \tag{2}$$

provided also that $a \leq 1$. If $a > 1$, the inequality can go either way.

Case 1. $0 < a \leq 1$. We can assume that $x \geq y \geq z$. Then

$$3x^2 + x^3 \geq 3a + a^3 \geq 3z + z^3,$$

so that $x \geq a \geq z$. Solving for y in (1) and substituting in (2), we obtain

$$x^2(1 + z - z^2) + x(z^2 + z - p) + z^2 - pz + p \geq 0, \tag{3}$$

where $p = 3a + a^3$. Inequality (3) will be valid if the discriminant is less than or equal to zero; that is, if

$$(z^2 + z - p)^2 \leq 4(1 + z - z^2)(z^2 - pz + p). \tag{4}$$

Inequality (4) can be rewritten as

$$p(4 - p) + z(1 - z)\,(2p(1 - 2z) + z(5z + 3)) \geq 0.$$

Since $p \leq 4$ and $z \leq 1$, the inequality clearly holds for $z \leq 1/2$. For $z > 1/2$, we have

$$2p(1 - 2z) + z(5z + 3) \geq 8(1 - 2z) + z(5z + 3) = (1 - z)(8 - 5z) \geq 0.$$

It now follows that the only time equality can occur is if $a = 1$ and $x = y = z = 1$, or $a = 1$ and $z = 0$, $x = y = 2$.

Case 2. Here $p > 4$. Let $y = z = \varepsilon$. Then $x \approx p/2\varepsilon$, so that (2) holds. Now let $z = 0$, $x = y$. Then $x = y = \sqrt{p}$, so that the inequality sign in (2) is reversed.

The real numbers x, y, z, t satisfy the equalities $x + y + z + t = 0$ and $x^2 + y^2 + z^2 + t^2 = 1$. Prove that $-1 \leq xy + yz + zt + tx \leq 0$.

Solution

First, we have:

$$xy + yz + zt + tx = (x + z)(y + t) = -(x + z)^2 \leq 0$$

(since $y + t = -(x + z)$). Then,

$$|xy + yz + zt + tx| \leq (x^2 + y^2 + z^2 + t^2)^{1/2}(y^2 + z^2 + t^2 + x^2)^{1/2} = 1$$

(by the Cauchy-Schwarz inequality). The conclusion follows.

Remark. Equality $xy + yz + zt + tx = 0$ holds if and only if $x + z = y + t = 0$. Therefore, inequality $xy + yz + zt + tx \leq 0$ becomes an equality for the quadruplets $(x, y, z, t) = (a, b, -a, -b)$ where a, b are real numbers such that $a^2 + b^2 = \frac{1}{2}$.

If equality $xy + yz + zt + tx = -1$ holds, then we have equality in the Cauchy-Schwarz inequality so that (x, y, z, t) and (y, z, t, x) are proportional. Since at least one of the numbers x, y, z, t must be non-zero, this leads to $x = y = z = t$ or $y = -x$, $z = x$, $t = -x$. The first case is incompatible with the hypotheses, and the second case provides only the quadruplets:

$$\left(\frac{1}{2}, -\frac{1}{2}, \frac{1}{2}, -\frac{1}{2}\right) \quad \text{and} \quad \left(-\frac{1}{2}, \frac{1}{2}, -\frac{1}{2}, \frac{1}{2}\right).$$

Conversely, these two quadruplets satisfy $xy + yz + zt + tx = -1$ and we may conclude: $xy + yz + zt + tx = -1$ holds if and only if $(x, y, z, t) = (\frac{1}{2}, -\frac{1}{2}, \frac{1}{2}, -\frac{1}{2})$ or $(-\frac{1}{2}, \frac{1}{2}, -\frac{1}{2}, \frac{1}{2})$.

Natural numbers k, n are given such that $1 < k < n$. Solve the system of n equations

$$x_i^3 \cdot (x_i^2 + x_{i+1}^2 + \cdots + x_{i+k-1}^2) = x_{i-1}^2 \quad \text{for} \quad 1 \le i \le n,$$

with n real unknowns x_1, x_2, ..., x_n. Note: $x_0 = x_n$, $x_{n+1} = x_1$, $x_{n+2} = x_2$, and so on.

Solution

We prove that there are two solutions

$$x_1 = x_2 = \cdots = x_n = 0$$

and

$$x_1 = x_2 = \cdots = x_n = \frac{1}{\sqrt[3]{k}}.$$

First, suppose that $(x_1, \ldots, x_n)$ is a solution.

First Case. There is $i \in \{1, \ldots, n\}$ such that $x_i = 0$.

From the cyclic symmetry, we suppose, without loss of generality, that $x_1 = 0$. Then

$$x_1^3(x_1^2 + \cdots + x_k^2) = 0 = x_n^2.$$

Thus, $x_n = 0$.

Continuing, we deduce that $x_i = 0$ for all i.

Second Case. $x_i \ne 0$ for all i.

Since $x_i^3(x_i^2 + \cdots + x_{i+k-1}^2) = x_{i-1}^2$, we deduce that $x_i > 0$ for all i. From cyclic symmetry, without loss of generality, we have

$$x_1 = \min \{x_i : i = 1, 2, \ldots, n\}.$$

Then $x_1^3 \le x_2^3$ and $x_1^2 \le x_{k+1}^2$. Thus,

$$x_1^2 + x_2^2 + \cdots + x_k^2 \le x_2^2 + x_3^2 + \cdots + x_{k+1}^2.$$

We deduce that

$$
\begin{aligned}
x_n^2 &= x_1^3(x_1^2 + \cdots + x_k^2) \\
&\le x_2^3(x_2^2 + \cdots + x_{k+1}^2) = x_1^2.
\end{aligned}
$$

Thus, $x_n \le x_1$.

What we have actually shown is that if $x_i = \min\{x_1, \ldots, x_n\}$, then $x_{i-1} = x_i \ (= \min\{x_1, \ldots, x_n\})$. An easy induction leads to $x_1 = x_2 = \cdots = x_n$.

Let a denote the common value. Then $a^3(ka^2) = a^2$; that is, $a = \frac{1}{\sqrt[3]{k}}$.

Conversely, it is easy to verify that $(0, 0, \ldots, 0)$ and $\left(\frac{1}{\sqrt[3]{k}}, \ldots, \frac{1}{\sqrt[3]{k}}\right)$ are solutions.

Show that there do not exist non-negative integers k and m such that $k! + 48 = 48(k+1)^m$.

Solution

Suppose the given equation holds for some non-negative integers k and m. Then $48 \mid k!$. Since $48 = 2^4 \times 3$, we must have $k \geq 6$. If $k = 6$ or 7, the equation becomes $16 = 7^m$ or $106 = 8^m$, respectively. Clearly, neither is possible. Hence, $k \geq 8$ and the given equation can be rewritten as

$$3 \times 5 \times 7 \times 8 \times \cdots \times (k-1) \times k + 1 = (k+1)^m. \qquad (1)$$

Suppose $k + 1$ is a composite. Then it has a prime divisor q. Since $q \leq k$, we have $q \mid k!$, which implies $q \mid 48$. Since $k \geq 8$, the left side of (1) is odd and thus, q must be odd. Hence, $q = 3$, which is clearly impossible in view of (1), since $3 \nmid 1$.

Therefore, $k + 1 = p$ is a prime. Then $k! = (p-1)! \equiv -1 \pmod{p}$ by Wilson's Theorem; that is, $p \mid k! + 1$.

Rewriting the given equation as $k! + 1 + 47 = 48p^m$, we infer that $p \mid 47$ and so $p = 47$. Then we have $46! + 48 = 48 \times 47^m$ or $46! = 48(47^m - 1)$. Since the prime divisors of $46!$ include $5, 7, 11$ which are all coprime with 48, we have $47^m \equiv 1 \pmod{5, 7, 11}$. Now, straightforward checking reveals that $\text{ord}_5(47) = 4$ (that is, the least positive integer n such that $47^n \equiv 1 \pmod 5$ is $n = 4$), $\text{ord}_7(47) = 6$ and $\text{ord}_{11}(47) = 5$. Hence, m is divisible by $\text{lcm}\{4, 6, 5\}$; that is, $60 \mid m$. So $m \geq 60$ and we have $48 \times 47^m \geq 48 \times 47^{60}$. Clearly, $48 \times 47^{60} > 46! + 48$ and we have a contradiction.

Given real numbers $0 = x_1 < x_2 < \cdots < x_{2n} < x_{2n+1} = 1$ with $x_{i+1} - x_i \leq h$ for $1 \leq i \leq 2n$, show that

$$\frac{1-h}{2} < \sum_{i=1}^{n} x_{2i}(x_{2i+1} - x_{2i-1}) \leq \frac{1+h}{2}.$$

Solution

The problem is equivalent to showing that

$$\left| \sum_{i=1}^{n} x_{2i}(x_{2i+1} - x_{2i-1}) - \frac{1}{2} \right| < \frac{h}{2}.$$

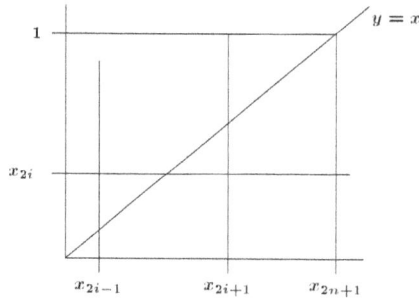

Now $\sum_{i=1}^{n} x_{2i}(x_{2i+1} - x_{2i-1}) - \frac{1}{2}$ is the difference between the area of the rectangles formed by the four lines $x = x_{2i-1}$, $x = x_{2i+1}$, $y = 0$ and $y = x_{2i}$ and the triangle formed by the three lines $x = 0$, $y = 1$, $x = y$. The area contained in the rectangles but not in the triangle (respectively contained in the triangle but not in the rectangles) is a union of triangles of total base less than 1 and height $\leq h$. Hence, we have the required inequality.

For which ordered pairs of positive real numbers (a, b) is the limit of every sequence (x_n) satisfying the condition

$$\lim_{n \to \infty} (ax_{n+1} - bx_n) = 0 \qquad (1)$$

zero?

Solution

If $b > a$, then, for $x_n = (b/a)^n$, we have $ax_{n+1} - bx_n = 0$, but $\lim_{n \to \infty} (b/a)^n = \infty$.

If $b = a$, we have the well-known example:

$$x_n = 1 + \frac{1}{2} + \cdots + \frac{1}{n},$$

for which we have $\lim_{n \to \infty} (x_{n+1} - x_n) = 0$, but $\{x_n\}$ does not converge to a finite limit.

Let us assume that $b < a$. We shall prove that

$$\underline{\lim} x_n = \overline{\lim} x_n = 0.$$

Denote by m (resp. M), the $\underline{\lim}$ (resp. $\overline{\lim}$) of $\{x_n\}$. By (1), we have $M \leq \frac{b}{a} m$, and since $m \leq M$, we deduce that $M \leq \frac{b}{a} M$, and consequently that $m, M \leq 0$. Similarly, by (1), we have that $M \frac{b}{a} \leq m$, and since $M \geq m$, we deduce that $m \geq \frac{b}{a} m$, and consequently that $m, M \geq 0$; whence, $m = M = 0$ and $\lim_{n \to \infty} x_n = 0$.

Let $a_1, a_2, \ldots, a_n$ be real numbers and s, a non-negative real number, such that

(i) $a_1 \leq a_2 \leq \cdots \leq a_n$;   (ii) $a_1 + a_2 + \cdots + a_n = 0$;

(iii) $|a_1| + |a_2| + \cdots + |a_n| = s$.

Prove that

$$a_n - a_1 \geq \frac{2s}{n}.$$

Solution

The result is clear when $s = 0$, so we will suppose $s > 0$. This implies that at least one of the a_i's is non-zero. Since $a_1 + a_2 + \cdots + a_n = 0$, the a_i's cannot all be non-negative, or all be non-positive. Thus:

$$a_1 = \min(a_i) < 0 \quad \text{and} \quad a_n = \max(a_i) > 0.$$

There exists $k \in \{1, 2, \ldots, n-1\}$ such that $a_1 \leq a_2 \leq \cdots \leq a_k \leq 0 < a_{k+1} \leq \cdots \leq a_n$. Then $a_{k+1} + \cdots + a_n = -(a_1 + a_2 + \cdots + a_k) = |a_1| + |a_2| + \cdots + |a_k| = s - (|a_{k+1}| + \cdots + |a_n|) = s - (a_{k+1} + \cdots + a_n)$. Hence, $a_{k+1} + \cdots + a_n = \frac{s}{2} = -(a_1 + a_2 + \cdots + a_k)$. For $i \in \{1, 2, \ldots, k\}$, $j \in \{k+1, \ldots, n\}$, we have: $a_j - a_i \leq a_n - a_1 (= \delta$, say) so that $a_n - a_1 \leq \delta, a_n - a_2 \leq \delta, \ldots, a_n - a_k \leq \delta$. Adding up, we get:
$ka_n + \frac{s}{2} \leq k\delta$. Substituting successively $a_{n-1}, \ldots, a_{k+1}$ for a_n, we get similarly:

$$ka_{n-1} + \frac{s}{2} \leq k\delta, \ldots, ka_{k+1} + \frac{s}{2} \leq k\delta.$$

Adding up again, we obtain: $k(a_n + \cdots a_{k+1}) + (n-k)\frac{s}{2} \leq (n-k)k\delta$ or $k\frac{s}{2} + (n-k)\frac{s}{2} \leq (n-k)k\delta$, which leads to $\delta \geq \frac{ns}{2k(n-k)}$. But $(n-k) + k = n$; hence, $k(n-k) \leq \frac{n^2}{4}$ and $\frac{ns}{2k(n-k)} \geq \frac{2s}{n}$. Thus, $\delta \geq \frac{2s}{n}$.

Let $n \geq 2$ be a fixed natural number and let $a_1, a_2, \ldots, a_n$ be positive numbers whose sum equals 1.

(a) Prove the inequality

$$2\sum_{i<j} x_i x_j \leq \frac{n-2}{n-1} + \sum_{i=1}^{n} \frac{a_i x_i^2}{1-a_i}$$

for any positive numbers $x_1, x_2, \ldots, x_n$ summing to 1.

(b) Determine all n-tuples of positive numbers $x_1, x_2, \ldots, x_n$ summing to 1 for which equality holds.

Solution

(a) Replacing a_i in the numerator by $(a_i - 1) + 1$ and simplifying, we get

$$1 \leq (n-1) \sum \frac{x_i^2}{1-a_i} = \left(\sum(1-a_i)\right)\left(\sum \frac{x_i^2}{1-a_i}\right),$$

where the sums here and subsequently are over $i = 1$ to n. By Cauchy's inequality, the right hand side is $\geq (\sum x_i)^2 = 1$.

(b) There is equality if and only if $x_i = k(1-a_i)$ where $k = \frac{1}{n-1}$.

The sum of three integers a, b and c is 0. Prove that $2a^4 + 2b^4 + 2c^4$ is the square of an integer.

Solution

Let $p(x) = x^3 + sx^2 + qx + r$ be a third degree polynomial having a, b, and c as roots.

According to Viète's Theorem $s = -(a + b + c) = 0$. Then

$$a^3 + qa + r = 0$$
$$b^3 + qb + r = 0$$
$$c^3 + qc + r = 0.$$

Upon multiplying each equation by $2a$, $2b$ and $2c$, respectively, and adding we have

$$2a^4 + 2b^4 + 2c^4 + 2q(a^2 + b^2 + c^2) = 0$$

(the term with r vanishing since $a + b + c = 0$). However,

$$a^2 + b^2 + c^2 = (a + b + c)^2 - 2(ab + bc + ac)$$
$$= -2q.$$

Thus,

$$2a^4 + 2b^4 + 2c^4 - 4q^2 = 0$$

and

$$2a^4 + 2b^4 + 2c^4 = (2q)^2$$

Prove that if

$$a_0^{a_1} = a_1^{a_2} = \cdots = a_{1995}^{a_{1996}} = a_{1996}^{a_0}, \quad a_1 \in \mathbb{R}^*,$$

then

$$a_0 = a_1 = \cdots = a_{1996}.$$

Solution

I think that the a_i are supposed to be non-negative real numbers, and give that solution.

If $a_0, \ldots, a_{1996}$ are non-negative real numbers such that

$$a_0^{a_1} = \cdots = a_{1996}^{a_0}, \quad \text{then } a_0 = a_1 = \cdots = a_{1996}.$$

We first prove that $a_i \neq 0$ for all i.

Indeed, we are given $a_1 \neq 0$. Then, $a_1^{a_2} = a_0^{a_1} \neq 0$. Thus, $a_0 \neq 0$. An easy induction yields $a_i \neq 0$ for all i.

Next we show that either $a_i = 1$ for all i, or $0 < a_i < 1$ for all i or $1 < a_i$ for all i.

First suppose $a_i = 1$. Then $a_i^{a_{i+1}} = 1 = a_{i+2}^{a_{i+3}}$ (indices read modulo 1997). Now $a_{i+2} \neq 1$ entails $a_{i+3} = 0$, contrary to the last claim. It follows then that $a_i = 1$ for some i implies that $a_j = 1$ for all $j = 0, 1, \ldots, 1996$. Thus, we may suppose that $a_i \neq 0, 1$ for all $i = 0, 1, \ldots, 1996$. Suppose, for a contradiction, that $0 < a_i < 1$ for some i and $1 < a_j$ for some value of j. Then, we may suppose, without loss of generality, that $0 < a_i < 1$ and $1 < a_{i+1}$. But then $1 > a_i^{a_{i+1}} = a_{i+1}^{a_{i+2}}$ gives $a_{i+2} < 0$, contrary to the hypothesis. The claim now follows.

To complete the proof we distinguish two cases:

Case 1. $0 < a_i < 1$ for all $i = 0, 1, \ldots, 1996$.

Now first suppose $a_0 < a_1$. From $a_1^{a_2} = a_0^{a_1} < a_1^{a_1}$ and the fact that a_1^x is monotone decreasing, we obtain $a_1 < a_2$.

From this we get $a_0 < a_1 < a_i < \cdots < a_{1996} < a_{1997} = a_0$, a contradiction. The assumption that $a_1 < a_0$ similarly leads to a contradiction, completing this case.

Case 2. $1 < a_i$ for all i.

Suppose for a contradiction, that $a_0 < a_1$. Then $a_1^{a_2} = a_0^{a_1} < a_1^{a_1}$. It follows that $a_2 < a_1$. Thus, $a_1^{a_2} = a_2^{a_3} < a_1^{a_3}$, and we obtain $a_2 < a_3$.

By an easy induction we get $a_{2p} < a_{2p+1}$ for all p (subscripts are read modulo 1997).

But $a_{1998} = a_1$ and $a_{1999} = a_2$, giving $a_2 > a_1$, a contradiction. Thus, we have $a_0 \geq a_1$.

Next suppose $a_0 > a_1$. By similar reasoning we obtain $a_{2p} > a_{2p+1}$ for all p, leading to a similar contradiction.

Thus, $a_0 = a_1$.

From the cyclic symmetry of the assumptions we get $a_i = a_{i+1}$ for all i; that is, $a_0 = a_1 = a_2 = \cdots = a_{1996}$.

Prove that for positive real numbers a and b

$$2 \cdot \sqrt{a} + 3 \cdot \sqrt[3]{b} \geq 5 \cdot \sqrt[5]{ab}.$$

Solution

Set $a_1 = \sqrt{a}$, $a_2 = \sqrt{a}$, $a_3 = \sqrt[3]{b}$, $a_4 = \sqrt[3]{b}$, $a_5 = \sqrt[3]{b}$. Then, by the A.M.–G.M. inequality, we have

$$\sqrt[5]{ab} = \sqrt[5]{a_1 \cdot a_2 \cdot a_3 \cdot a_4 \cdot a_5} \leq \frac{a_1 + a_2 + a_3 + a_4 + a_5}{5} = \frac{2}{5}\sqrt{a} + \frac{3}{5}\sqrt[3]{b}$$

and the result follows.

Find the biggest value of the difference $x - y$ if $2 \cdot (x^2 + y^2) = x + y$.

Solution

More generally: Let $p > 0$ and $q > 1$. If the real numbers x and y satisfy the equation

$$p(|x - y|^q + |x + y|^q) = x + y, \qquad (1)$$

then there holds the sharp inequality

$$x - y \leq (pq)^{1/(1-q)}(q - 1)^{1/q}. \qquad (2)$$

Proof. The function $f(t) = t/p - t^q$, $t \geq 0$, has first derivative $f'(t) = \frac{1}{p} - qt^{q-1}$, $t > 0$. Let $t_0 = (pq)^{1/(1-q)}$. From $f'(t) > 0$ if $0 < t < t_0$, and $f'(t) < 0$ if $t > t_0$, it follows that $f(t) \leq f(t_0)$ for all $t \geq 0$; that is,

$$\frac{t}{p} - t^q \leq (pq)^{q/(1-q)}(q - 1), \quad t \geq 0.$$

Suppose that the real numbers x and y satisfy (1). Then

$$|x - y|^q = \frac{x + y}{p} - |x + y|^q \leq \frac{|x + y|}{p} - |x + y|^q$$
$$\leq (pq)^{q/(1-q)}(q - 1).$$

Hence,

$$x - y \leq |x - y| \leq (pq)^{1/(1-q)}(q - 1)^{1/q},$$

proving (2). It is easily verified that

$$x = \frac{1}{2}(pq)^{1/(1-q)}(1 + (q - 1)^{1/q})$$

and

$$y = \frac{1}{2}(pq)^{1/(1-q)}(1 - (q - 1)^{1/q})$$

satisfy (1), and that there is equality in (2) for these values. Thus, (2) is sharp.

With $p = 1$ and $q = 2$, (1) becomes $2(x^2 + y^2) = x + y$. Then, by (2), there holds the sharp inequality $x - y \leq \frac{1}{2}$. Hence, $\frac{1}{2}$ is the biggest value asked for.

Solve the equation $x^{1996} - 1996x^{1995} + \cdots + 1 = 0$ (the coefficients in front of $x, \ldots, x^{1994}$ are unknown), if it is known that its roots are positive real numbers.

Solution

Let $x_1, x_2, \ldots, x_{1996}$ be the roots of the given polynomial. By hypothesis, $x_k > 0$ for all k in $\{1, 2, \ldots, 1996\}$ and from the known coefficients: $x_1 + x_2 + \cdots + x_{1996} = 1996$ and $x_1 \cdot x_2 \cdots x_{1996} = 1$. It follows that:

$$\frac{x_1 + x_2 + \cdots + x_{1996}}{1996} = 1 = \sqrt[1996]{x_1 \cdot x_2 \cdots x_{1996}}.$$

Therefore, we are in the case where the A.M.–G.M. inequality is actually an equality. As it is well known, this means that $x_1 = x_2 = \cdots = x_{1996}$ from which $x_1 = x_2 = \cdots = x_{1996} = 1$ immediately follows. Thus, the solution of the given equation is 1 with multiplicity 1996.

Determine all triples (x, y, z), satisfying

$$xy = z, \qquad (1)$$
$$xz = y, \qquad (2)$$
$$yz = x. \qquad (3)$$

Solution

If (x, y, z) is a solution, then, multiplying, we have $(xyz)^2 = xyz$. Thus, $xyz = 0$ or $xyz = 1$.

Case 1. If $xyz = 0$ then, for example, $z = 0$.
From (2) we get $y = 0$, and from (3) we obtain $x = 0$. Conversely, it is easy to see that $(0, 0, 0)$ is a solution.

Case 2. If $xyz = 1$ then $z = \frac{1}{xy}$.
From (1), we deduce $z^2 = 1$. Thus, $z \in \{-1, 1\}$.

In the same way, $x \in \{-1, 1\}$ and $y \in \{-1, 1\}$. Moreover, since $xyz = 1$, the number of -1's in (x, y, z) is even. This leads to

$$(-1, -1, 1), \quad (-1, 1, -1), \quad (1, -1, -1), \quad (1, 1, 1).$$

Conversely, it is easy to see that these triples are solutions. Then,

$$S = \{(0, 0, 0), (1, 1, 1), (-1, -1, 1), (-1, 1, -1), (1, -1, -1)\}.$$

Serge was solving the equation $f(19x - 96/x) = 0$ and found 11 different solutions. Prove that if he tried hard he would be able to find at least one more solution.

Solution

Note first that $x = 0$ is not a solution. If $r \neq 0$ is a solution, then so is $t = -\frac{96}{19r}$, since

$$f\left(19t - \frac{96}{t}\right) = f\left(-\frac{96}{r} + 19r\right) = 0.$$

Since $r = \frac{-96}{19r}$ is impossible, different r's will correspond to different t's. Therefore, the number of solutions must be even (if it is finite) and the conclusion easily follows.

It is known about real numbers $a_1, \ldots, a_{n+1}; b_1, \ldots, b_n$ that
$0 \le b_k \le 1$ $(k = 1, \ldots, n)$ and $a_1 \ge a_2 \ge \cdots \ge a_{n+1} = 0$. Prove the inequality:

$$\sum_{i=1}^{n} a_k b_k \le \sum_{k=1}^{[\sum_{j=1}^{n} b_j]+1} a_k .$$

Solution

Let $m = \left[\sum_{j=1}^{n} b_j\right]$. If $m = 0$, then $0 \le \sum_{j=1}^{n} b_j < 1$, so that

$$\sum_{k=1}^{n} a_k b_k \le a_1 \sum_{j=1}^{n} b_j \le a_1 = \sum_{k=1}^{m+1} a_k .$$

If $m = n$, then $b_k = 1$ $(k = 1, \ldots, n)$ and the inequality to be proved is obvious. Also, the case $n = 1$ is immediate, so that, in the following, we will suppose $n \ge 2$ and $1 \le m < n$.

Let

$s(1)$ be the first integer ≥ 2 such that $\displaystyle\sum_{j=1}^{s(1)} b_j \ge 1$,

$s(2)$ be the first integer $> s(1)$ such that $\displaystyle\sum_{j=1}^{s(2)} b_j \ge 2$,

$\ldots\ldots\ldots\ldots\ldots$

$s(m)$ be the first integer $> s(m-1)$ such that $\displaystyle\sum_{j=1}^{s(m)} b_j \ge m$.

Note that $s(k) \ge k + 1$, so that $a_{s(k)} \le a_{k+1}$ $(k = 1, \ldots, m)$. Now,

$$\sum_{k=1}^{n} a_k b_k = \sum_{k=1}^{n} c_k \text{ where } c_k = (a_k - a_{k+1}) \sum_{j=1}^{k} b_j \quad (\text{using } a_{n+1} = 0)$$

$$= A_0 + A_1 + \cdots + A_m$$

with

$$A_0 = \sum_{k=1}^{s(1)-1} c_k, \ A_1 = \sum_{k=s(1)}^{s(2)-1} c_k, \ \ldots, \ A_m = \sum_{k=s(m)}^{n} c_k .$$

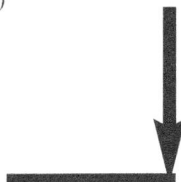

Let $t \in \{0, 1, \ldots, m\}$. Then for $s(t) \leq k \leq s(t+1) - 1$ (defining $s(0) = 1$, $s(m+1) = n+1$), we have $\sum_{j=1}^{k} b_j \leq t+1$ so that

$$A_t \leq (t+1) \sum_{k=s(t)}^{s(t+1)-1} c_k = (t+1)(a_{s(t)} - a_{s(t+1)}).$$

Thus, $A_0 \leq a_1 - a_{s(1)}$, $A_1 \leq 2(a_{s(1)} - a_{s(2)})$, $A_2 \leq 3(a_{s(2)} - a_{s(3)})$, $\ldots$, $A_m \leq (m+1)(a_{s(m)} - a_{n+1}) = (m+1)a_{s(m)}$, and, adding:

$$\sum_{k=1}^{n} a_k b_k \leq a_1 + a_{s(1)} + \cdots + a_{s(m)} \leq a_1 + a_2 + \cdots + a_{m+1}.$$

Let $n = 2^{13} \cdot 3^{11} \cdot 5^7$. Find the number of divisors of n^2 which are less than n and are not divisors of n.

Solution

The divisors of n^2 are the integers $2^a \cdot 3^b \cdot 5^c$ where the integers a, b, c satisfy: $0 \le a \le 26$, $0 \le b \le 22$, $0 \le c \le 14$, so that there are $N = 27 \times 23 \times 15$ divisors of n^2 altogether.

Let $N = 2K + 1$. We can partition the set of all divisors of n^2 as follows:

$$\{n\}, \{d_1, d_{K+1}\}, \{d_2, d_{K+2}\}, \ldots, \{d_K, d_{2K}\}$$

where, for $i = 1, 2, \ldots, K$:

$$d_i < n, \quad d_{K+i} > n \quad \text{and} \quad d_i \cdot d_{K+i} = n^2.$$

From this, we see that exactly K divisors of n^2 are less than n. Moreover, the divisors of n (n excepted) are all among these divisors. Hence, the number we seek is $K - (14 \times 12 \times 8 - 1)$; that is

$$\frac{27 \times 23 \times 15 - 1}{2} - 14 \times 12 \times 8 + 1 = 3314.$$

Distinct square trinomials $f(x)$ and $g(x)$ have leading coefficient equal to one. It is known that $f(-12) + f(2000) + f(4000) = g(-12) + g(2000) + g(4000)$. Find all the real values of x which satisfy the equation $f(x) = g(x)$.

Solution

Let $f(x) = x^2 + ax + b$ and $g(x) = x^2 + cx + d$ be two distinct square trinomials.

From the given inequality we get

$$1996a + b = 1996c + d.$$

Thus, we have $a \neq c$ and $b \neq d$.

Therefore, $x = 1996$ is the single root of the equation $f(x) = g(x)$.

Prove the equality

$$\frac{1}{666} + \frac{1}{667} + \cdots + \frac{1}{1996} = 1 + \frac{2}{2 \cdot 3 \cdot 4} + \frac{2}{5 \cdot 6 \cdot 7} + \cdots + \frac{2}{1994 \cdot 1995 \cdot 1996}.$$

Solution

We prove that, in general, for all natural numbers n,

$$1 + \frac{2}{2 \cdot 3 \cdot 4} + \frac{2}{5 \cdot 6 \cdot 7} + \cdots + \frac{2}{(3n-1)(3n)(3n+1)}$$
$$= \frac{1}{n+1} + \frac{1}{n+2} + \cdots + \frac{1}{3n+1}. \tag{1}$$

The given equality is the special case when $n = 665$.

To prove (1), let S_1 and S_2 denote the left and right hand sides of (1), respectively. By simple partial fractions, we find that

$$\frac{2}{(3k-1)(3k)(3k+1)} = \frac{1}{3k-1} - \frac{2}{3k} + \frac{1}{3k+1} \quad \text{for all } k \geq 1.$$

Hence,

$$
\begin{aligned}
S_1 &= 1 + \sum_{k=1}^{n} \frac{2}{(3k-1)(3k)(3k+1)} \\
&= 1 + \sum_{k=1}^{n} \left(\frac{1}{3k-1} - \frac{2}{3k} + \frac{1}{3k+1} \right) \\
&= 1 + \sum_{k=1}^{n} \left(\frac{1}{3k-1} + \frac{1}{3k} + \frac{1}{3k+1} \right) - 3\sum_{k=1}^{n} \frac{1}{3k} \\
&= 1 + \sum_{k=2}^{3n+1} \frac{1}{k} - \sum_{k=1}^{n} \frac{1}{k} \\
&= \sum_{k=n+1}^{3n+1} \frac{1}{k} \\
&= S_2.
\end{aligned}
$$

Prove that the product of the roots of the equation

$$\sqrt{1996} \cdot x^{\log_{1996} x} = x^6$$

is an integer number and find the last four digits of this number.

Solution

Let $a = \log_{1996} x$. Since $\log_{1996}(\sqrt{1996}) = \frac{1}{2}$, the given equation is equivalent to

$$\frac{1}{2} + a^2 = 6a \quad \text{(by applying } \log_{1996} \text{ to each side).}$$

Denote by a_1 and a_2 the two roots of this equation. Then the roots of the given equation are $x_1 = 1996^{a_1}$ and $x_2 = 1996^{a_2}$ whose product is $1996^{a_1 + a_2} = 1996^6$. Thus, the product $x_1 x_2$ is the integer 1996^6.

Now $1996^6 = (2000 - 4)^6 = 2^6(2 - 1000)^6 = 2^6(2^6 - 6 \cdot 2^5 \cdot 1000 + \cdots)$ [dots represent terms all multiple of 10000, of no influence on the last four digits]. Hence,

$$
\begin{aligned}
1996^6 &\equiv 2^6(2^6 - 192000) \pmod{10000} \\
&\equiv 64(64 + 8000) \pmod{10000} \\
&\equiv 6096 \pmod{10000},
\end{aligned}
$$

so that the last four digits of 1996^6 are 6 0 9 6.

Prove that for all natural numbers $m \geq 2$ and $n \geq 2$ the smallest among the numbers $\sqrt[m]{m}$ and $\sqrt[n]{n}$ does not exceed the number $\sqrt[3]{3}$.

Solution

For positive integers k, let $x_k = k^{1/k}$. First note that

$$\sqrt[3]{3} > 1.44. \tag{1}$$

Case 1. $m = n$.

From (1), we have $x_1 < x_2 < x_3$.

Suppose that, for a fixed integer $p \geq 3$, we have $x_p \leq x_3$. Then $p^3 \leq 3^p$ (since $(x \mapsto x^{3p})$ is non-decreasing). It follows that:

$$
\begin{aligned}
3^{p+1} &\geq 3p^3 \\
&= p^3 + 3p^2 + 3p + (p-3)p^2 + (p^2 - 3)p \\
&> p^3 + 3p^2 + 3p + 1 \quad \text{(since } p \geq 3\text{)} \\
&= (p+1)^3.
\end{aligned}
$$

Then, we have $x_{p+1} \leq x_3$.

By induction, we then have $x_p \leq x_3$ for all $p \geq 3$. Thus,

$$x_p \leq x_3 \quad \text{for all positive integers } p.$$

Case 2. $m \neq n$.

With no loss of generality, we may suppose that $m < n$.

Then

$$
\begin{aligned}
\sqrt[m]{m} &< \sqrt[n]{n} \quad ((x \mapsto \sqrt[n]{x}) \text{ is increasing}) \\
&\leq \sqrt[3]{3} \quad \text{from case 1.}
\end{aligned}
$$

Then, for all positive integers m, n, the smallest among $\sqrt[n]{n}$ and $\sqrt[m]{m}$ does not exceed the number $\sqrt[3]{3}$.

Prove the inequality $2^{a_1} + 2^{a_2} + \cdots + 2^{a_{1996}} \le 1995 + 2^{a_1 + a_2 + \cdots + a_{1996}}$
for any real non-positive numbers $a_1, a_2, \ldots, a_{1996}$.

First we prove the following lemma.

Lemma. Let $f : X \to \mathrm{R}$ be a function for $X \subset \mathrm{R}$.

Suppose that $f(x + y) = f(x)f(y)$ for all $x, y \in X$. For $a_1, a_2, \ldots,$
$a_n \in X$, let $s_i = \sum_{j=1}^{i} a_j$ for $1 \le i \le n$. Then we have

$$f(s_n) + n - 1 = \sum_{i=1}^{n-1} (f(s_i) - 1)(f(a_{i+1}) - 1) + \sum_{i=1}^{n} f(a_i)$$

for $n \ge 2$.

Proof. We have

$$
\begin{aligned}
&\sum_{i=1}^{n-1} (f(s_i) - 1)(f(a_{i+1}) - 1) + \sum_{i=1}^{n} f(a_i) \\
&= \sum_{i=1}^{n-1} f(s_i)f(a_{i+1}) - \sum_{i=1}^{n-1} f(s_i) - \sum_{i=1}^{n-1} f(a_{i+1}) + \sum_{i=1}^{n-1} 1 + \sum_{i=1}^{n} f(a_i) \\
&= \sum_{i=1}^{n-1} f(s_i + a_{i+1}) - \sum_{i=1}^{n-1} f(s_i) - \sum_{i=2}^{n} f(a_i) + \sum_{i=1}^{n} f(a_i) + n - 1 \\
&= \sum_{i=1}^{n-1} f(s_{i+1}) - \sum_{i=1}^{n-1} f(s_i) - \sum_{i=2}^{n} f(a_i) + \sum_{i=1}^{n} f(a_i) + (n - 1) \\
&= \sum_{i=2}^{n} f(s_i) - \sum_{i=1}^{n-1} f(s_i) - \sum_{i=2}^{n} f(a_i) + \sum_{i=1}^{n} f(a_i) + (n - 1) \\
&= f(s_n) - f(s_1) + f(a_1) + (n - 1) = f(s_n) + (n - 1),
\end{aligned}
$$

since $s_1 = a_1$.

Since $2^{x+y} = 2^x \cdot 2^y$ for all $x, y \in \mathrm{R}$, we have

$$2^{a_1 + \cdots + a_{1996}} + 1995 = \sum_{i=1}^{1995} \left(2^{\sum_{j=1}^{i} a_j} - 1 \right)(2^{a_{i+1}} - 1) + \sum_{i=1}^{1996} 2^{a_i}$$

from the lemma. Since $2^\alpha \le 1$ for all $\alpha \le 0$, we easily deduce that

$$\sum_{i=1}^{1996} 2^{a_i} \le 2^{a_1 + \cdots + a_{1996}} + 1995, \qquad \text{as desired}$$

Solve in real numbers the equation

$$2x^2 - 3x = 1 + 2x\sqrt{x^2 - 3x}.$$

Solution

The given equation

$$2x^2 - 3x = 1 + 2x\sqrt{x^2 - 3x}$$

is equivalent to

$$x^2 - 2x\sqrt{x^2 - 3x} + x^2 - 3x = 1,$$
$$\text{or} \quad (x - \sqrt{x^2 - 3x})^2 - 1 = 0,$$
$$\text{or} \quad (x - \sqrt{x^2 - 3x} + 1)(x - \sqrt{x^2 - 3x} - 1) = 0.$$

The equation $x - \sqrt{x^2 - 3x} - 1 = 0$ has no roots, whereas the equation $x - \sqrt{x^2 - 3x} + 1 = 0$ has the root $x = -\frac{1}{5} = -0.2$.

Let x, y and z be non-negative real numbers satisfying $x + y + z = 1$. Show that

$$x^2 y + y^2 z + z^2 x \leq \frac{4}{27},$$

and find when equality occurs.

Solution

For a more direct solution but not particularly elegant, we can assume without loss of generality that $x \geq y \geq z$ (if we changed the cyclic order of these, $x^2 y + y^2 z + z^2 x$ would be decreased). In terms of a homogeneous inequality, we want to show that

$$4(x + y + z)^3 \geq 27(x^2 y + y^2 z + z^2 x).$$

We now set $z = a$, $y = a+b$, and $x = a+b+c$ where $a, b, c \geq 0$. Expanding out, we get

$$27a^3 + 54ab + 27a^2 c + 36ab^2 + 9ac^2 + 36abc + (5b + 4c)(b - c)^2 \geq 0$$

and which is now evident. There is equality, if and only if, $a = 0$ and $b = c$.

The sequence $\{a_n\}$, $n \geq 0$, is such that $a_0 = 1$, $a_{499} = 0$ and for $n \geq 1$, $a_{n+1} = 2a_1a_n - a_{n-1}$.

(a) Prove that $|a_1| \leq 1$.

(b) Find a_{1996}.

Solution

(a) Suppose on the contrary that $|a_1| > 1$.

The relation $a_{n+1} = 2a_1a_n - a_{n-1}$ for $n \geq 1$ leads to the characteristic equation $r^2 = 2a_1r - 1$. This equation has two distinct real solutions:

$$r_1 = a_1 + \sqrt{a_1^2 - 1} \quad \text{and} \quad r_2 = a_1 - \sqrt{a_1^2 - 1}.$$

Hence, a_n is given by $a_n = ur_1^n + vr_2^n$, where $u + v = a_0 = 1$ and $ur_1 + vr_2 = a_1$. Solving, we get $u = v = \frac{1}{2}$ so that

$$a_n = \frac{r_1^n + r_2^n}{2} \quad \text{for all } n \geq 0.$$

Since $r_1 > 0$, $r_2 > 0$, this last result contradicts $a_{499} = 0$. Therefore, $|a_1| \leq 1$.

(b) $|a_1| \leq 1$ enables one to set $a_1 = \cos\theta$. Then, an easy induction shows that $a_n = \cos(n\theta)$ for all $n \geq 0$. In particular $a_{499} = \cos(499\theta) = 0$. From $1996 = 4 \times 499$ and the formula $\cos 4t = 8\cos^4 t - 8\cos^2 t + 1$ (obtained for instance by expressing a_4 in terms of $a_1 = \cos(t)$), we get:

$$a_{1996} = 8\cos^4(499\theta) - 8\cos^2(499\theta) + 1 = 1.$$

Prove that the equation $a^2 + b^2 = c^2 + 3$ has infinitely many integer solutions (a, b, c).

Solution

We show more generally that "the equation $a^2 + b^2 = c^2 + n$ has infinitely many integer solutions (a, b, c) for any $n \in \mathbb{Z}$."

Case 1. n is even ($n = 2k$ for some $k \in \mathbb{Z}$)

$$n = 2k = (2t(t+1)-k)^2 + (2t+1)^2 - (2t(t+1)-k+1)^2 \quad \text{for all } t \in \mathbb{Z}.$$

Case 2. n is odd ($n = 2k - 1$ for some $k \in \mathbb{Z}$)

$$n = 2k-1 = ((1-2kt^2)k)^2 + (2kt)^2 - ((1-2kt^2)k - 1)^2 \quad \text{for all } t \in \mathbb{Z}$$

Prove that there are no integers m and n such that

$$419m^2 + 95mn + 2000n^2 = 1995.$$

Solution

Suppose integers m, n exist such that

$$419m^2 + 95mn + 2000n^2 = 1995.$$

Then, since 95, 2000 and 1995 are all divisible by 5, it follows that $5 \mid m^2 \implies 5 \mid m$. Now put $m = 5M$ and we have

$$(419)(25)M^2 + (95)(25)Mn + 2000n^2 = 1995.$$

Now the left-hand side is divisible by 25 and the right-hand side is not. The contradiction establishes that no integers m and n exist satisfying the given equation.

Suppose that $a_1, a_2, a_3, \ldots, a_n$ are the numbers $1, 2, 3, \ldots, n$ but written in any order. Prove that

$$(a_1 - 1)^2 + (a_2 - 2)^2 + (a_3 - 3)^2 + \cdots + (a_n - n)^2$$

is always even.

Solution

Let $S = \sum_{i=1}^{n}(a_i - i)^2$. Then

$$S = \sum_{i=1}^{n} a_i^2 + \sum_{i=1}^{n} i^2 - 2\sum_{i=1}^{n} i a_i.$$

Since $(a_1, \ldots, a_n)$ is a permutation of $(1, \ldots, n)$, we have

$$\sum_{i=1}^{n} a_i^2 = \sum_{i=1}^{n} i^2.$$

Thus, $S = 2\left(\sum_{i=1}^{n} i^2 - \sum_{i=1}^{n} i a_i\right)$ is even.

Find all pairs (m, n) of natural numbers with $m < n$ such that $m^2 + 1$ is a multiple of n and $n^2 + 1$ is a multiple of m.

Solution

We will prove that the solutions are the pairs of the form (f_{2n-1}, f_{2n+1}) for $n \in \mathbb{N}$, where $\{f_k\}$ is the Fibonacci sequence ($f_0 = 0$, $f_1 = 1$ and $f_{n+2} = f_n + f_{n+1}$).

Let (m, n) be such that $m, n \in \mathbb{N}^*$, $m < n$, and $m^2 + 1$ is a multiple of n and $n^2 + 1$ is a multiple of m.

If $m = 1$ then $n \geq 2$ and n divides 2. Thus, $n = 2$.

Conversely, $(1, 2)$ is a solution.

Claim: (m, n) is a solution with $m > 1$ if and only if $\left(\frac{m^2+1}{n}, m \right)$ is a solution.

Proof of the Claim: First note that $(m, (m^2 + 1)) = 1$ and $(n, (n^2 + 1)) = 1$. It follows that

(m, n) is a solution

$\Longleftrightarrow$ $\dfrac{(m^2 + 1)(n^2 + 1)}{mn}$ is an integer

$\Longleftrightarrow$ $\dfrac{m^2 + n^2 + 1}{mn}$ is an integer

$\Longleftrightarrow$ there exists $k \in \mathbb{N}^*$ such that $m^2 + n^2 + 1 = kmn$

$\Longleftrightarrow$ there exists $k \in \mathbb{N}^*$ such that n is a solution of the equation $(E_k) : X^2 - kmX + m^2 + 1 = 0$.

In the same way: (m, n) is a solution $\Longleftrightarrow$ there exists $k \in \mathbb{N}^*$ such that m is a solution of the equation $X^2 - knX + n^2 + 1 = 0$.

Since (E_k) is a quadratic equation with integer coefficients and leading coefficient equal to 1, one of its solutions is an integer if and only if the other is as well. Moreover n is a solution of (E_k) if and only if $\frac{m^2+1}{n}$ is a solution of (E_k) (from the product of the roots of (E_k)).

It follows that (m, n) is a solution $\Longleftrightarrow$ there exists $k \in \mathbb{N}^*$ such that $\frac{m^2+1}{n}$ is a solution of (E_k).

- If $m = 1$ then $n = 2$ and $\frac{m^2+1}{n} = m$.
- If $m > 1$ then, since $m + 1 \leq n$, we have

$$mn \geq m(m + 1) > m^2 + 1.$$

Thus, $\frac{m^2+1}{n} < m$.

We deduce that (m, n) is a solution with $m > 1 \iff \left(\frac{m^2+1}{n}, m\right)$ is a solution, and the claim is proved.

Let (m, n) be a solution, with $m > 1$. From the claim, we may find another solution (m', n') with $m' < n' = m < n$. Repeating this process, we construct a sequence (m_i, n_i) of solutions, and the sequence $\{n_i\}$ is a strictly decreasing sequence of positive integers. This sequence has to be finite. But the only way to stop the process is to obtain $m_i = 1$ (and then $n_i = 2$) for some k.

It follows that each solution of the problem is generated from $(1, 2)$ by using a finite number of applications of the function $(m, n) \to \left(n, \frac{n^2+1}{m}\right)$.

Conversely, if $m_1 = 1$ and $n_1 = 2$, and for $n \in \mathbb{N}^*$,

$$\begin{cases} m_{i+1} &=& n_i, \\ n_{i+1} &=& \frac{n_i^2+1}{m_i}. \end{cases} \tag{1}$$

From the claim, we know that (m_i, n_i) is a solution for all $i \geq 1$.

Then, the solutions of the problem are pairs (m_i, n_i) defined by (1), for all $i \geq 2$: $n_{i+1} = \frac{n_i^2+1}{m_i} = \frac{n_i^2+1}{n_{i-1}}$; that is

$$n_{i+1} n_{i-1} - 1 = n_i^2. \tag{2}$$

It looks like a well-known property of the Fibonacci sequence

$$f_n^2 = f_{n+1} f_{n-1} + (-1)^{n-1}$$

where $f_0 = 0$, $f_1 = 1$, $f_{n+2} = f_n + f_{n+1}$. It is also well known that for all $n \geq 0$

$$f_n = \frac{1}{\sqrt{5}} \left(\varphi^n - \left(-\frac{1}{\varphi}\right)^n\right) \quad \text{where} \quad \varphi = \frac{1+\sqrt{5}}{2} \tag{3}$$

Using (3) it is not difficult to see that f_{2k+1} satisfies (2). Since $f_1 = m_1$ (if we define $n_0 = m_1$ we then have $n_0 = f_1$), and $f_3 = n_1$, $f_5 = n_2$, it follows that for all $i \geq 1$

$$n_i = f_{2i+1}$$

and then

$$m_i = n_{i-1} = f_{2i-1},$$

And we are done.

Let a be a real number such that $0 < a \le 1$ and $a \le a_j \le \frac{1}{a}$, for $j = 1, 2, \ldots, 1996$. Show that for any non-negative real numbers λ_j ($j = 1, 2, \ldots, 1996$), with

$$\sum_{j=1}^{1996} \lambda_j = 1,$$

one has

$$\left(\sum_{i=1}^{1996} \lambda_i a_i \right) \left(\sum_{j=1}^{1996} \lambda_j a_j^{-1} \right) \le \frac{1}{4} \left(a + \frac{1}{a} \right)^2.$$

Solution

Let $n = 1996$ (or n be any positive integer, actually). From the AM GM Inequality,

$$\left(\sum_{i=1}^{n} \lambda_i a_i \right)^{1/2} \left(\sum_{j=1}^{n} \lambda_j a_j^{-1} \right)^{1/2} \le \frac{1}{2} \left(\sum_{i=1}^{n} \lambda_i a_i + \sum_{j=1}^{n} \lambda_j a_j^{-1} \right)$$

$$= \frac{1}{2} \left(\sum_{j=1}^{n} \lambda_j \left(a_j + \frac{1}{a_j} \right) \right).$$

Now, if t satisfies $a \le t \le \frac{1}{a}$, then $\left(t + \frac{1}{t} \right) - \left(a + \frac{1}{a} \right) = (t - a) \left(1 - \frac{1}{at} \right) \le 0$, so that $t + \frac{1}{t} \le a + \frac{1}{a}$. It follows that $a_j + \frac{1}{a_j} \le a + \frac{1}{a}$ ($j = 1, 2, \ldots, n$) and, since $\lambda_j \ge 0$, $\sum_{j=1}^{n} \lambda_j \left(a_j + \frac{1}{a_j} \right) \le \left(a + \frac{1}{a} \right) \sum_{j=1}^{n} \lambda_j = a + \frac{1}{a}$.

From this, we get

$$\left(\sum_{i=1}^{n} \lambda_i a_i \right)^{1/2} \left(\sum_{j=1}^{n} \lambda_j a_j^{-1} \right)^{1/2} \le \frac{1}{2} \left(a + \frac{1}{a} \right)$$

and the required result by squaring both sides.

Generalization: Replacing the hypothesis on the a_i by $0 < m \le a_i \le M$, we get the following inequality:

$$\left(\sum_{i=1}^{n} \lambda_i a_i \right) \left(\sum_{j=1}^{n} \lambda_j a_j^{-1} \right) \le \frac{1}{4} \frac{(m + M)^2}{mM}.$$

It suffices to remark that $\sqrt{\frac{m}{M}} \le \frac{a_i}{\sqrt{mM}} \le \sqrt{\frac{M}{m}}$ and apply the result above with $a = \sqrt{\frac{m}{M}}$ and $\frac{a_i}{\sqrt{mM}}$ instead of a_i.

Show that for any real numbers a_3, a_4, ..., a_{85}, the roots of the equation

$$a_{85}x^{85} + a_{84}x^{84} + \cdots + a_3x^3 + 3x^2 + 2x + 1 = 0$$

are not real.

Solution

Let $P(x) = a_{85}x^{85} + \cdots + a_3x^3 + 3x^2 + 2x + 1$. Since $P(0) = 1$, then 0 is not a root of P.

Let r_1, ..., r_{85} be the complex roots of P.

For $i = 1, \ldots, 85$, denote $s_i = \frac{1}{r_i}$. Then the s_i's are the complex roots of the polynomial $Q(y) = y^{85} + 2y^{84} + 3y^{83} + a_3y^{82} + \cdots + a_{84}y + a_{85}$. It follows that

$$\sum_{i=1}^{85} s_i = -2 \quad \text{and} \quad \sum_{i<j} s_i s_j = 3.$$

Then

$$\sum_{i=1}^{85} s_i^2 = \left(\sum_{i=1}^{85} s_i\right)^2 - 2\sum_{i<j} s_i s_j = -2 < 0.$$

Thus, the s_i's are not all real, and then the r_i's are not all real.

Is there any solution of the equation

$$\lfloor x \rfloor + \lfloor 2x \rfloor + \lfloor 4x \rfloor + \lfloor 8x \rfloor + \lfloor 16x \rfloor + \lfloor 32x \rfloor = 12345?$$

($\lfloor x \rfloor$ denotes the greatest integer which does not exceed x.)

Solution

The equation

$$\lfloor x \rfloor + \lfloor 2x \rfloor + \lfloor 4x \rfloor + \lfloor 8x \rfloor + \lfloor 16x \rfloor + \lfloor 32x \rfloor = 12345 \qquad (1)$$

has no real solution. Suppose, for a contradiction, that $x \in \mathbb{R}$ satisfies (1).

Then $x > 0$ and we may write

$$x = N + \frac{a}{2} + \frac{b}{4} + \frac{c}{8} + \frac{d}{16} + \frac{e}{32} + f$$

where N is a non-negative integer,

$$a, b, c, d, e \in \{0, 1\} \quad \text{and} \quad f \in \left[0, \tfrac{1}{32}\right).$$

From (1) we obtain

$$63N + 31a + 15b + 7c + 3d + e = 12345.$$

Thus,

$$63N \leq 12345 \leq 63N + 31 + 15 + 7 + 3 + 1.$$

That is,

$$\frac{12288}{63} \leq N \leq \frac{12345}{63}.$$

Then, $195 < N < 196$, which is impossible if N is supposed to be an integer. Thus, (1) has no solution, as claimed.

Determine all pairs of numbers λ_1, $\lambda_2 \in \mathbb{R}$ for which every solution of the equation

$$(x + i\lambda_1)^n + (x + i\lambda_2)^n = 0$$

is real. Find the solutions.

Solution

Let x be any complex number solution. Then, $(x+i\lambda_1)^n = -(x+i\lambda_2)^n$ so that

$$|x + i\lambda_1|^n = |x + i\lambda_2|^n \quad \text{and} \quad |x + i\lambda_1| = |x + i\lambda_2|. \qquad (1)$$

Denote by A_1, A_2 the points in the complex plane corresponding to $-i\lambda_1$, $-i\lambda_2$ respectively. The relation (1) means that $MA_1 = MA_2$, where M corresponds to x. Thus, M is on the perpendicular bisector Δ of segment $A_1 A_2$. Note that Δ is a line perpendicular to the imaginary axis. It follows that every solution is a real number if and only if Δ coincides with the real axis; that is, $\lambda_1 + \lambda_2 = 0$.

Conversely, suppose $\lambda_1 = -\lambda_2 = \lambda \in \mathbb{R}$. The given equation becomes $(x+i\lambda)^n = -(x-i\lambda)^n$. If $\lambda = 0$, then $x = 0$ is the only solution. Assuming now that $\lambda \neq 0$, our equation is equivalent to:

$$\left(\frac{x + i\lambda}{x - i\lambda} \right)^n = -1 \quad \text{or} \quad \frac{x + i\lambda}{x - i\lambda} = u_k$$

where $u_k = \exp(i(\frac{\pi}{n} + \frac{2k\pi}{n}))$ for $k = 0, 1, \ldots, n-1$. This gives $x = i\lambda \frac{u_k+1}{u_k-1}$ or, by an easy computation $x = \lambda \cot \left(\frac{(2k+1)\pi}{2n} \right)$.

Thus, the solutions are the n real numbers $x_k = \lambda \cot \left(\frac{(2k+1)\pi}{2n} \right)$, ($k = 0, 1, \ldots, n-1$).

Find all pairs of consecutive integers the difference of whose cubes is a full square.

Solution

Let a, b be two integers.

$$(a+1)^3 - a^3 = b^2 \iff 3a^2 + 3a + 1 = b^2 \qquad (1)$$
$$\iff 3(4a^2 + 4a + 1) + 1 = 4b^2$$
$$\iff (2b)^2 - 3(2a+1)^2 = 1$$
$$\iff (2b, 2a+1) \quad \text{is a solution}$$

of Pell's equation

$$X^2 - 3Y^2 = 1. \qquad (2)$$

The minimal non-trivial solution of (2) is $(2, 1)$. It is then well known that the solutions of (2) are the pairs $(\pm x_n, \pm y_n)$ where $x_0 = 1$, $y_0 = 0$, and for all $n \geq 0$

$$\begin{cases} x_{n+1} = 2x_n + 3y_n \\ y_{n+1} = 2y_n + x_n \end{cases}.$$

But we want only those with x_n even and y_n odd.

It is easy to see that if x_n is even and y_n is odd then x_{n+1} is odd and y_{n+1} is even, and then x_{n+2} is even and y_{n+2} is odd.

Thus, since $x_1 = 2$ and $y_1 = 1$, we consider only the pairs (x_{2n+1}, y_{2n+1}). Since, for all $n \geq 0$

$$\begin{cases} x_{n+2} = 7x_n + 12y_n \\ y_{n+2} = 4x_n + 7y_n \end{cases}.$$

Then, the solutions of (1) are the pairs (a, b) of the form

$$\left(\frac{-1 \pm V_n}{2}, \pm \frac{U_n}{2} \right) \quad \text{where} \quad U_1 = 2, V_1 = 1$$

and for all $n \geq 1$

$$\begin{cases} U_{n+1} = 7U_n + 12V_n \\ V_{n+1} = 4U_n + 7V_n \end{cases}.$$

For example, we first note that $(a+1)^3 - a^3 = b^2$ if and only if $(-a)^3 - (-a-1)^3 = b^2$. Then we give only the first positive values of a and b.

n	U_n	V_n	a	b
1	2	1	0	1
2	26	16	7	13
3	362	209	104	181
4	5042	2911	1455	2521
5	70226	40545	20272	35113
$\vdots$				

Prove that for every natural number $n \geq 3$ there exist two sets $A = \{x_1, x_2, \ldots, x_n\}$ and $B = \{y_1, y_2, \ldots, y_n\}$ such that

(a) $A \cap B = \emptyset$,

(b) $x_1 + x_2 + \cdots + x_n = y_1 + y_2 + \cdots + y_n$,

(c) $x_1^2 + x_2^2 + \cdots + x_n^2 = y_1^2 + y_2^2 + \cdots + y_n^2$.

Solution

We can choose $(n - 1)$ positive numbers $a_1, \ldots, a_{n-1}$ such that $a_1 < a_2 < \cdots < a_{n-1}$. Then we have

$$-\sum_{i=1}^{n-1} a_i < -a_{n-1} < \cdots < -a_2 < -a_1 < 0 < a_1 < a_2 < \cdots < a_{n-1} < \sum_{j=1}^{n-1} a_i.$$

It follows that

$$\{a_1, \ldots, a_{n-1}, -\sum_{i=1}^{n-1} a_i\} \cap \{-a_1, \ldots, -a_{n-1}, \sum_{i=1}^{n-1} a_i\} = \emptyset.$$

Now, let $x_n = -\sum_{i=1}^{n-1} a_i$, $y_n = \sum_{i=1}^{n-1} a_i$ and $x_i = a_i$, $y_i = -a_i$ for $1 \leq i \leq n - 1$. Then we get $\sum_{i=1}^{n} x_i = 0 = \sum_{i=1}^{n} y_i$ and

$$\sum_{i=1}^{n} x_i^2 = \sum_{i=1}^{n-1} a_i^2 + \left(\sum_{i=1}^{n-1} a_i\right)^2 = \sum_{i=1}^{n} y_i^2.$$

Prove that for any natural number n

$$\lceil \sqrt{n} + \sqrt{n+1} + \sqrt{n+2} \rceil = \lceil \sqrt{9n+8} \rceil.$$

Solution

For $n = 0, 1, 2$
it is easy to see that the relation is true. Now, for $n \geq 3$, we have

$$n(n+1)(n+2) > \left(n + \frac{8}{9}\right)^3$$

(consider the function $(x-1)x(x+1) - (x - \frac{1}{9})^3$).

By the AM-GM inequality, we have

$$\frac{\sqrt{n} + \sqrt{n+1} + \sqrt{n+2}}{3} > \sqrt[3]{\sqrt{n}\sqrt{n+1}\sqrt{n+2}}$$

$$> \sqrt{n + \frac{8}{9}}.$$

Hence,

$$\sqrt{n} + \sqrt{n+1} + \sqrt{n+2} > \sqrt{9n+8}.$$

On the other hand,

$$\frac{\sqrt{n} + \sqrt{n+1} + \sqrt{n+2}}{3} < \sqrt{\frac{n+n+1+n+2}{3}}.$$

Hence,

$$\sqrt{n} + \sqrt{n+1} + \sqrt{n+2} < \sqrt{9n+9}.$$

Consequently,

$$\lceil \sqrt{n} + \sqrt{n+1} + \sqrt{n+2} \rceil = \lceil \sqrt{9n+8} \rceil.$$

Prove the following inequality

$$(xy + xz + yz) \left(\frac{1}{(x+y)^2} + \frac{1}{(y+z)^2} + \frac{1}{(x+z)^2} \right) \geq \frac{9}{4}$$

for positive real numbers x, y, z.

Solution

Reducing to the same denominator, the inequality to be proved is equivalent to

$$A + B + C \geq 0$$

where

$$A =: \sum_{\text{symmetric}} \left(4x^5 y - x^4 y^2 - 3x^3 y^3 \right)$$

$$B =: \sum_{\text{symmetric}} \left(4xy^5 - x^2 y^4 - 3x^3 y^3 \right)$$

$$C =: \sum_{\text{symmetric}} \left(2x^4 yz - 2x^3 y^2 z - 2x^3 yz^2 + 2x^2 y^2 z^2 \right)$$

and $\sum\limits_{\text{symmetric}}$ runs over all six permutations of x, y, z.

Thanks to Shur's inequality,

$$x(x-y)(x-z) + y(y-z)(y-x) + z(z-x)(z-y) \geq 0,$$

we have then by multiplying by $2xyz$:

$$C = \sum_{\text{symmetric}} \left(2x^4 yz - 2x^3 y^2 z - 2x^3 yz^2 + 2x^2 y^2 z^2 \right) \geq 0.$$

On the other hand, by the rearrangement inequality, we have

$$\sum_{\text{symmetric}} x^5 y \geq \sum_{\text{symmetric}} x^4 y^2, \qquad \sum_{\text{symmetric}} x^5 y \geq \sum_{\text{symmetric}} x^3 y^3,$$

and hence,

$$A = \sum_{\text{symmetric}} \left(4x^5 y - x^4 y^2 - 3x^3 y^3 \right) \geq 0,$$

$$B = \sum_{\text{symmetric}} \left(4xy^5 - x^2 y^4 - 3x^3 y^3 \right) \geq 0.$$

Let k be a positive integer. Prove that there are infinitely many perfect squares in the arithmetic progression $\{n \times 2^k - 7\}_{n \geq 1}$.

Solution

We first show, by induction on m, that, for every m, there exists a positive number a_m for which $a_m^2 \equiv -7 \pmod{2^m}$.

Note that $a_m = 1$ satisfies the conditions for $m \leq 3$. Inductively, let us suppose that $a_m^2 \equiv -7 \pmod{2^m}$. Then $a_m^2 \equiv -7 \pmod{2^{m+1}}$, or $a_m^2 \equiv 2^m - 7 \pmod{2^{m+1}}$.

If $a_m^2 \equiv -7 \pmod{2^{m+1}}$, then we can put $a_{m+1} = a_m$. If $a_m^2 \equiv 2^m - 7 \pmod{2^{m+1}}$, then we put $a_{m+1} = a_m + 2^{m-1}$, and note that

$$
\begin{aligned}
a_{m+1}^2 &= (a_m + 2^{m-1})^2 = a_m^2 + 2^m a_m + 2^{2m-2} \\
&\equiv a_m^2 + 2^m a_m \pmod{2^{m+1}} \\
&\equiv a_m^2 + 2^m \pmod{2^{m+1}} \equiv -7 \pmod{2^{m+1}}
\end{aligned}
$$

completing the induction step.

Now since $a_m^2 \geq 2^m - 7$, the sequence (a_m) is unbounded and thus, takes on infinitely many values; that is, there are infinitely many numbers m for which one can find an associated positive number n_m such that $n_m 2^m - 7$ $(= a_m^2)$ is a perfect square.

Now, given our (fixed) number k, we simply consider the infinitely many m for which $m \geq k$, and note that $(n_m \cdot 2^{m-k}) 2^k - 7$ is a perfect square.

Let a, b, c be positive real numbers. Find all real numbers x, y, z such that

$$x + y + z = a + b + c$$

$$4xyz - (a^2x + b^2y + c^2z) = abc.$$

Solution

The second equation is equivalent to

$$4 = \frac{a^2}{yz} + \frac{b^2}{zx} + \frac{c^2}{xy} + \frac{abc}{xyz}$$

and also to

$$4 = x_1^2 + y_1^2 + z_1^2 + x_1 y_1 z_1,$$

where

$$0 < x_1 = \frac{a}{\sqrt{yz}} < 2, \qquad 0 < y_1 = \frac{b}{\sqrt{zx}} < 2,$$

$$0 < z_1 = \frac{c}{\sqrt{xy}} < 2.$$

Setting $x_1 = 2\sin u$, $0 < u < \frac{\pi}{2}$, and $y_1 = 2\sin v$, $0 < v < \frac{\pi}{2}$, we have

$$4 = 4\sin^2 u + 4\sin^2 v + z_1^2 + 4\sin u \cdot \sin v \cdot z_1.$$

Hence,

$$z_1 + 2\sin u \cdot \sin v = 2\cos u \cdot \cos v,$$

and then,

$$z_1 = 2(\cos u \cdot \cos v - \sin u \cdot \sin v) \quad (= 2\cos(u+v)).$$

Thus,

$$a = 2\sqrt{yz}\sin u,$$
$$b = 2\sqrt{zx}\sin v,$$
$$c = 2\sqrt{xy}(\cos u \cdot \cos v - \sin u \cdot \sin v).$$

From $x + y + z = a + b + c$, we get

$$\left(\sqrt{x}\cos v - \sqrt{y}\cos u\right)^2 + \left(\sqrt{x}\sin v + \sqrt{y}\sin u - \sqrt{z}\right)^2 = 0$$

which implies

$$\sqrt{z} = \sqrt{x}\sin v + \sqrt{y}\sin u = \sqrt{x}\frac{y_1}{2} + \sqrt{y}\frac{x_1}{2}.$$

Therefore,

$$\sqrt{z} = \sqrt{x} \cdot \frac{b}{2\sqrt{zx}} + \sqrt{y} \cdot \frac{a}{2\sqrt{yz}},$$

and thus, $z = \frac{a+b}{2}$. Similarly, $x = \frac{a+b}{2}$, $y = \frac{c+a}{2}$.

The triple

$$(x, y, z) = \left(\frac{b+c}{2}, \frac{c+a}{2}, \frac{a+b}{2}\right)$$

is the unique solution.

The numbers x, y and $\dfrac{x^2 + y^2 + 6}{xy}$ are positive integers. Prove that

$\dfrac{x^2 + y^2 + 6}{xy}$ is a perfect cube.

Solution

Suppose $\dfrac{x^2 + y^2 + 6}{xy} = k$ for some positive integers x, y and k. We prove that necessarily $k = 8$.

Consider

$$\frac{x^2 + y^2 + 6}{xy} = k \qquad (1)$$

as a Diophantine equation in two variables x and y. Let (a, b) denote the solution of (1) with the *least* positive a value. We first show that $a = 1$. Due to symmetry, we may assume that $a \leq b$. Note first that a and b must both be odd since they clearly must have the same parity and if they are both even, then modulo 4, $a^2 + b^2 + 6 \equiv 2$ while $kab \equiv 0$. If $a = b$, then from $k = 2 + \frac{6}{a^2}$ we deduce immediately that $a = 1$. If $a < b$ then $b \geq a + 1$.

Since

$$\frac{(ka - b)^2 + a^2 + 6}{(ka - b)a} = \frac{k^2 a^2 - 2kab + kab}{ka^2 - ab} = k$$

and

$$ka - b = \frac{a^2 + b^2 + 6}{b} - b = \frac{a^2 + 6}{b} > 0,$$

we see that $(ka - b, a)$ is also a solution of (1) in natural numbers. Note that $a^2 + 6 - ab \leq a^2 + 6 - a(a+1) = 6 - a < 0$ if $a > 6$. Thus, $ka - b = \frac{a^2 + 6}{b} < a$ if $a > 6$, contradicting the minimality of a. Hence, $a \leq 6$. It remains to show that $a \neq 3, 5$. If $a = 3$, then we get $b^2 + 15 = 3kb$, which implies that 3 divides b and thus, $b^2 + 15 \equiv 6$, while $3kb \equiv 0 \pmod{9}$. If $a = 5$, then we get $b^2 + 31 = 5kb$. Since b is an integer and 31 is a prime we deduce from the relation between roots and coefficients that $5k = 32$, which is impossible. Therefore, we conclude that $a = 1$, which implies $k = \dfrac{b^2 + 7}{b} = b + \dfrac{7}{b}$. Hence, $b = 7$ and $k = 8$, which is a cube.

Find all sequences of integers $x_1, x_2, \ldots, x_{1997}$ such that

$$\sum_{k=1}^{1997} 2^{k-1}(x_k)^{1997} = 1996 \prod_{k=1}^{1997} x_k .$$

Solution

We make the following key observation: if $(x_1, x_2, \ldots, x_{1997})$ satisfies the property

$$\sum_{k=1}^{1997} 2^{k-1}(x_k)^{1997} = 1996 \prod_{k=1}^{1997} x_k , \qquad (1)$$

then x_1 is even.

Substituting $2y_1$ for x_1 in (1) and dividing both sides by 2, we get:

$$(x_2)^{1997} + 2(x_3)^{1997} + \cdots + 2^{1995}(x_{1997})^{1997} + 2^{1996}(y_1)^{1997}$$
$$= 1996 \left(\prod_{k=2}^{1997} x_k \right) y_1 .$$

This means that the sequence $(x_2, \ldots, x_{1997}, y_1)$ also satisfies (1).

Iterating, we obtain successively:

x_2 is even, $x_2 = 2y_2$ and $(x_3, \ldots, x_{1997}, y_1, y_2)$ satisfies (1)

$\ldots \ldots \ldots \ldots \ldots \ldots$

x_{1997} is even, $x_{1997} = 2y_{1997}$, and $(y_1, y_2, \ldots, y_{1997})$ satisfies (1)

Thus, we have obtained the following result: if $(x_1, x_2, \ldots, x_{1997})$ satisfies (1), then the x_k's are even and $\left(\frac{x_1}{2}, \frac{x_2}{2}, \ldots, \frac{x_{1997}}{2} \right)$ also satisfies (1). It follows that all the x_k's are 0. Indeed, if $x_j \neq 0$ (say), then we might write $x_j = 2^r \cdot z$ with $r \geq 1$ and z odd; iterating the previous process r times would yield a sequence satisfying (1) and containing the odd term z, which is impossible, as we have seen. Conversely, the null sequence obviously satisfies (1) so that we may conclude that the only sequence satisfying (1) is the null sequence.

Let a_1, a_2, ..., a_n be arbitrary real numbers and b_1, b_2, ..., b_n real numbers satisfying the condition $1 \geq b_1 \geq b_2 \geq \cdots \geq b_n \geq 0$. Prove that there is a positive integer $k \leq n$ for which the inequality $|a_1b_1 + a_2b_2 + \cdots + a_nb_n| \leq |a_1 + a_2 + \cdots + a_k|$ holds.

Solution

Let $A_1 = a_1$, $A_2 = a_1 + a_2$, ..., $A_n = a_1 + a_2 + \cdots + a_n$ and $M = \max(|A_1|, |A_2|, \ldots, |A_n|)$. We have to prove that

$$|a_1b_1 + a_2b_2 + \cdots + a_nb_n| \leq M.$$

Noticing that

$$
\begin{aligned}
& a_1b_1 + a_2b_2 + \cdots + a_nb_n \\
&= A_1b_1 + (A_2 - A_1)b_2 + \cdots + (A_n - A_{n-1})b_n \\
&= A_1(b_1 - b_2) + A_2(b_2 - b_3) + \cdots + A_{n-1}(b_{n-1} - b_n) + A_nb_n,
\end{aligned}
$$

we obtain:

$$
\begin{aligned}
& |a_1b_1 + a_2b_2 + \cdots + a_nb_n| \\
&\leq |A_1||b_1 - b_2| + |A_2||b_2 - b_3| + \cdots + |A_{n-1}||b_{n-1} - b_n| + |A_n||b_n| \\
&\leq M(|b_1 - b_2| + |b_2 - b_3| + \cdots + |b_{n-1} - b_n| + |b_n|) \\
&= M(b_1 - b_2 + b_2 - b_3 + \cdots + b_{n-1} - b_n + b_n) \\
&= Mb_1 \leq M, \quad \text{as desired.}
\end{aligned}
$$

Determine all numbers a, for which the equation

$$a3^x + 3^{-x} = 3$$

has a unique solution x.

Solution

We will prove that the required numbers a are those which satisfy $a \leq 0$ or $a = \dfrac{9}{4}$. Let a, x be real numbers. Let $f(x) = 3^{1-x} - 3^{-2x}$. Then $a3^x + 3^{-x} = 3 \iff f(x) = a$.

For all real numbers x : $f'(x) = \ln(3) \cdot 3^{-2x}(2 - 3^{1+x})$.

Let $x_0 = \dfrac{\ln\left(\frac{2}{3}\right)}{\ln(3)}$ (that is, $3^{1+x_0} = 2$). We then have

x	$-\infty$		x_0		$+\infty$
$f'(x)$		$+$	0	$-$	
$f(x)$	$-\infty$	$\nearrow$	$f(x_0)$	$\searrow$	0

and, since $3^{1+x_0} = 2$, we have $f(x_0) = 3 \cdot \dfrac{3}{2} - \left(\dfrac{3}{2}\right)^2 = \dfrac{9}{4}$.

From this, we easily deduce that the equation $f(x) = a$ has a unique solution if and only if $a = f(x_0) = \dfrac{9}{4}$ or $a \leq 0$. And we are done.

Find all triples (x, y, z) of integers satisfying the inequality:

$$x^2 + y^2 + z^2 + 3 \; < \; xy + 3y + 2z \, .$$

Solution

The inequality can be rewritten in the form

$$\left(x - \frac{y}{2}\right)^2 + \frac{3(y-2)^2}{4} + (z-1)^2 \; < \; 1 \, .$$

Hence, we have the following bounds:

$$-1 \; < \; z - 1 \; < \; 1, \quad \frac{-2}{\sqrt{3}} \; < \; y - 2 \; < \; \frac{2}{\sqrt{3}}, \quad -1 \; < \; x - \frac{y}{2} \; < \; 1 \, ,$$

so that we can only have $z = 1$, $y = 1$, 2, or 3 and $x = 0$, 1, or 2. Finally, the only solution is

$$(x, y, z) \; = \; (1, 2, 1) \, .$$

If the inequality sign $<$ in the given problem were to be changed to the equal sign $=$, we would have more solutions. Here we can only have $z = 0$, 1, or 2, $y = 1$, 2, or 3, and $x = 0$, 1, 2, or 3. Then the only solutions are $(x, y, z) = (0, 1, 1)$, $(0, 2, 1)$, $(1, 2, 0)$, $(1, 1, 1)$, $(1, 3, 1)$, $(1, 2, 2)$, $(2, 2, 1)$ and $(2, 3, 1)$.

Let $n \geq 3$. Suppose that the sequence $a_1, a_2, \ldots, a_n$ of positive real numbers satisfies $a_{i-1} + a_{i+1} = k_i a_i$, $\forall i = 1, 2, \ldots, n$, where each k_i is a positive integer, $a_0 = a_n$, $a_{n+1} = a_1$. Show that

$$2n \leq k_1 + k_2 + \cdots + k_n \leq 3n.$$

Solution

Since $k_i = \dfrac{a_{i-1} + a_{i+1}}{a_i} = \dfrac{a_{i-1}}{a_i} + \dfrac{a_{i+1}}{a_i}$ for all $i = 1, 2, \ldots, n$, we have

$$k_1 + k_2 + \cdots + k_n = \sum_{i=1}^{n} \left(\frac{a_i}{a_{i+1}} + \frac{a_{i+1}}{a_i} \right) \geq n \times 2 = 2n.$$

To prove the other inequality, we use induction on n: for $n = 3$ the result is easily verified. Assume that $k_1 + k_2 + \cdots + k_{n-1} \leq 3n - 3$ and let us prove the following

$$k_1 + k_2 + \cdots + k_n \leq 3n.$$

If all the a_i are equal, then it is true. Otherwise, there exists i such that $(a_i \geq a_{i-1}$ and $a_i > a_{i+1})$ or $(a_i > a_{i-1}$ and $a_i \geq a_{i+1})$. Hence, $a_{i-1} + a_{i+1} < 2a_i$ and thus, $k_i = 1$, and consequently the sequence $\{a_1, a_2, \ldots, a_{i-1}, a_{i+1}, \ldots, a_n\}$ satisfies the same condition with $\{k_1, k_2, \ldots, k_{i-2}, k_{i-1} - 1, k_{i+1} - 1, k_{i+2}, \ldots, k_n\}$. Since

$$k_1 + k_2 + \cdots + k_{i-2} + k_{i-1} - 1 + k_{i+1} - 1 + k_{i+2} + \cdots + k_n \leq 3n - 3,$$

then we obtain

$$k_1 + k_2 + \cdots + k_n \leq 3n - 3 + 2 + 1 = 3n.$$

(Brazil): Given a natural number $n \geq 2$, all the fractions of the form $\frac{1}{ab}$, with a and b natural numbers, coprime and such that

$$a < b \leq n, \qquad a + b > n,$$

are considered. Show that the sum of all these fractions equals $\frac{1}{2}$.

Solution

Let $\mathcal{I}_n$ be the set of all pairs (a, b) of coprime integers satisfying $1 \leq a < b \leq n$ and $a + b > n$. We observe:

- if $(a, b) \in \mathcal{I}_n$, then $(a, b) \in \mathcal{I}_{n+1}$ except when $a + b = n + 1$.
- if $(a, b) \in \mathcal{I}_{n+1}$, then $(a, b) \in \mathcal{I}_n$ except when $b = n + 1$.

From these observations, denoting by S_n the sum $\sum_{(a,b) \in \mathcal{I}_n} \frac{1}{ab}$, we get $S_{n+1} - S_n = s_1 - s_2$, where s_1 is the sum of all $\frac{1}{ab}$ with $(a, b) \in \mathcal{I}_{n+1}$ and $b = n + 1$ and s_2 is the sum of all $\frac{1}{ab}$ with $(a, b) \in \mathcal{I}_n$ and $a + b = n + 1$.

Using $\gcd(a, n + 1) = 1 \iff \gcd(n + 1 - a, n + 1) = 1$, we may write

$$
\begin{aligned}
s_1 &= \sum_{1 \leq a < n+1}^{*} \frac{1}{(n+1)a} \\
&= \sum_{1 \leq a < \frac{n+1}{2}}^{*} \frac{1}{n+1}\left(\frac{1}{a} + \frac{1}{n+1-a}\right) = \sum_{1 \leq a < \frac{n+1}{2}}^{*} \frac{1}{a(n+1-a)}
\end{aligned}
$$

where the $*$ indicates that we keep only terms for which $\gcd(a, n+1) = 1$.

Now, if $(a, b) \in \mathcal{I}_n$ and $a + b = n + 1$, we have $b = n + 1 - a$ and $a < \frac{n+1}{2}$ (because $n + 1 = a + b > 2a$), so that

$$
s_2 = \sum_{\gcd(a,n+1-a)=1}^{**} \frac{1}{a(n+1-a)} = \sum_{\gcd(a,n+1)=1}^{**} \frac{1}{a(n+1-a)}
$$

where the $**$ indicates that we keep only terms for which $1 \leq a < \dfrac{n+1}{2}$.

Thus, $s_2 = s_1$ and $S_{n+1} = S_n$. It follows that $S_n = S_1 = \frac{1}{2}$ for all n.

The sum of two of the roots of the equation

$$x^3 - 503x^2 + (a + 4)x - a = 0$$

is equal to 4. Determine the value of a.

Solution

We have $a = 1996$ — what a surprise!

Let α, β, γ denote the roots of $x^3 - 503x^2 + (a + 4)x - a = 0$ with $\alpha + \beta = 4$.

Since $\alpha + \beta + \gamma = 503$, we deduce that $\gamma = 499$. Then

$$499^3 - 503 \cdot 499^2 + (a + 4)499 - a = 0.$$

Thus, $a = 1996$ as claimed.

If a, b, c are positive real numbers, prove the inequality

$$a^2 + b^2 + c^2 - ab - bc - ca \geq 3(b - c)(a - b).$$

When is the "$=$" sign valid?

Solution

We put $X = a - b$ and $Y = b - c$, so that $X + Y = a - c$. Since

$$
\begin{aligned}
& 2(a^2 + b^2 + c^2 - ab - bc - ca) - 6(b - c)(a - b) \\
& = (a - b)^2 + (b - c)^2 + (c - a)^2 - 6(b - c)(a - b) \\
& = X^2 + Y^2 + (X + Y)^2 - 6XY \\
& = 2(X^2 + Y^2 - 2XY) \\
& = 2(X - Y)^2 \geq 0 \quad \text{(equality holds when } X = Y\text{)}.
\end{aligned}
$$

Thus, we have

$$2(a^2 + b^2 + c^2 - ab - bc - ca) \geq 6(b - c)(a - b).$$

Further, we obtain

$$a^2 + b^2 + c^2 - ab - bc - ca \geq 3(b - c)(a - b).$$

Equality holds when $X = Y$; that is $a - b = b - c$. This implies that $a + c = 2b$.

Let $a_1, a_2, \ldots$ be non-negative numbers which satisfy

$$a_{n+m} \leq a_n + a_m \quad (m, n \in \mathbb{N}).$$

Prove that

$$a_n \leq ma_1 + \left(\frac{n}{m} - 1\right) a_m$$

for all $n \geq m$.

Solution

From the hypothesis, we immediately get $a_{kn} \leq ka_n$ for all $k, n \in \mathbb{N}$; in particular $a_k \leq ka_1$ for all k.

We will show the desired property by induction on the quotient q in the division of n by m. Assume first that $q = 1$. In other words, $n = m + r$ where $r \in \{0, 1, \ldots, m-1\}$. We have to show that

$$a_{m+r} \leq ma_1 + \frac{r}{m}a_m. \tag{1}$$

We have $a_{m+r} \leq a_r + a_m \leq ra_1 + a_m$, but also $a_{m+r} \leq (m+r)a_1$, so that

$$ma_{m+r} = ra_{m+r} + (m-r)a_{m+r} \leq r^2a_1 + ra_m + (m-r)(m+r)a_1$$
$$= m^2a_1 + ra_m.$$

Thus, (1) follows immediately.

Suppose now that $a_{qm+r} \leq ma_1 + \left(\frac{qm+r}{m} - 1\right) a_m$ is true for an integer $q \geq 1$ and any $r \in \{0, 1, \ldots, m-1\}$. We prove that

$$a_{(q+1)m+r} \leq ma_1 + \left(\frac{(q+1)m+r}{m} - 1\right) a_m. \tag{2}$$

Indeed,

$$a_{(q+1)m+r}$$
$$= a_{qm+r+m} \leq a_m + a_{qm+r} \leq a_m + ma_1 + \left(\frac{qm+r}{m} - 1\right) a_m$$
$$= ma_1 + \frac{qm+r}{m}a_m = ma_1 + \left(\frac{(q+1)m+r}{m} - 1\right) a_m.$$

Hence, we get (2), and this completes the proof.

The positive integers x_1, x_2, x_3, x_4, x_5, x_6, x_7 satisfy the conditions:

$$x_6 = 144 \quad \text{and} \quad x_{n+3} = x_{n+2}(x_{n+1} + x_n) \quad \text{for} \quad n = 1, 2, 3, 4.$$

Compute x_7.

 Solution

If we substitute the equations for $n = 1$ and 2 into the equation for $n = 3$, we get

$$x_3(x_1 + x_2)(x_2 + x_3)(x_3 + x_4) = 144.$$

Now we can get lower bounds

$$\begin{aligned}
x_4 &= x_3(x_1 + x_2) \geq 2x_3, \\
x_5 &= x_4(x_2 + x_3) \geq 2x_3(1 + x_3) > 2x_3^2, \\
144 &\geq 2x_3(3x_3^2) = 6x_3^3.
\end{aligned}$$

Thus, x_3 is 1 or 2.

Suppose that $x_3 = 1$. We get

$$(x_1 + x_2)(1 + x_1 + x_2)(x_2 + 1) = 144$$

using the fact that $x_1 + x_2$ and $1 + x_1 + x_2$ are consecutive factors in

$$144 = (1 \times 2) \times 72 = (2 \times 3) \times 24 = (3 \times 4) \times 12 = (8 \times 9) \times 2.$$

An easy calculation shows us that the only possible value for $x_1 + x_2$ is 8. This gives $x_1 = 7$, $x_2 = 1$, $x_3 = 1$, $x_2 = 8$, $x_5 = 16$, $x_6 = 144$ and $x_7 = 3456$.

Suppose that $x_3 = 2$. We get

$$(x_1 + x_2)(1 + x_1 + x_2)(x_2 + 2) = 36.$$

The same ideas as above give us $x_1 = 1$, $x_2 = 2$, $x_3 = 2$, $x_4 = 6$, $x_5 = 18$, $x_6 = 144$, and $x_7 = 3456$.

Solve the following system of equations in real numbers x, y, z:

$$3(x^2 + y^2 + z^2) = 1 \qquad (1)$$
$$x^2 y^2 + y^2 z^2 + z^2 x^2 = xyz(x + y + z)^3 . \qquad (2)$$

Solution

Since all the terms in the sum of the left hand side of (2) are non-negative, we have $xyz(x + y + z)^3 \geq 0$ and hence $xyz(x + y + z) \geq 0$.

If $xyz(x + y + z) = 0$, equation (2) implies that $xy = yz = zx = 0$. From this and (1), it follows that exactly two of the numbers x, y, z are equal to zero.

If $x = y = 0$, substitution in (1) yields $z = \pm\frac{1}{\sqrt{3}}$.

Similarly, if $y = z = 0$ then $x = \pm\frac{1}{\sqrt{3}}$. If $z = x = 0$, then $y = \pm\frac{1}{\sqrt{3}}$.

If $xyz(x + y + z) > 0$, we consider the inequality

$$(u + v + w)^2 \leq 3(u^2 + v^2 + w^2) \qquad (3)$$

which is a special case of Cauchy's Inequality

$$(\alpha u + \beta v + \gamma w)^2 \leq (\alpha^2 + \beta^2 + \gamma^2)(u^2 + v^2 + w^2)$$

with $\alpha = \beta = \gamma = 1$.

By (3) applied with $u = xy$, $v = yz$, $w = zx$, we have

$$(xy + yz + zx)^2 \leq 3(x^2 y^2 + y^2 z^2 + z^2 x^2)$$

which simplifies to

$$\begin{aligned} xyz(x + y + z) &\leq x^2 y^2 + y^2 z^2 + z^2 x^2 \\ &= xyz(x + y + z)^3 \quad \text{(by (2).)} \end{aligned}$$

Dividing by the positive factor $xyz(x + y + z)$, we get

$$\begin{aligned} 1 &\leq (x + y + z)^2 \\ &\leq 3(x^2 + y^2 + z^2) \quad \text{(by (3))} \quad = 1 \quad \text{(by (1).)} \end{aligned}$$

Therefore,

$$(x + y + z)^2 = 3(x^2 + y^2 + z^2) ,$$

which is equivalent to

$$(x - y)^2 + (y - z)^2 + (z - x)^2 = 0 ,$$

and this holds if and only if $x = y = z$.

Solving for x, y and z, we immediately obtain

$$x = y = z = \frac{1}{3} \quad \text{or} \quad x = y = z = -\frac{1}{3} .$$

Thus, the general solution is constituted by the following triples (x, y, z):

$$\left(\frac{1}{\sqrt{3}}, 0, 0\right) , \left(-\frac{1}{\sqrt{3}}, 0, 0\right) , \left(0, \frac{1}{\sqrt{3}}, 0\right) , \left(0, -\frac{1}{\sqrt{3}}, 0\right) ,$$

$$\left(0, 0, \frac{1}{\sqrt{3}}\right) , \left(0, 0, -\frac{1}{\sqrt{3}}\right) , \left(\frac{1}{3}, \frac{1}{3}, \frac{1}{3}\right) , \left(-\frac{1}{3}, -\frac{1}{3}, -\frac{1}{3}\right) .$$

The sequence a_1, a_2, a_3, ... is defined by

$$a_1 = 0 \quad a_n = a_{\lfloor n/2 \rfloor} + (-1)^{n(n+1)/2} \quad \text{for} \quad n > 1 \,.$$

For every integer $k \geq 0$, find the number of all n such that

$$2^k \leq n < 2^{k+1}, \quad a_n = 0$$

($\lfloor n/2 \rfloor$ denotes the greatest integer not exceeding $n/2$).

Solution

For every integer k the number of n such that $2^k \leq n < 2^{k+1}$ and $a_n = 0$ is $\binom{k}{k/2}$ if k is even, and 0 if k is odd. It easily follows from the solutions of problem No. 8 proposed to the IMO but not used in 1996 (see [1998 : 466–470] and [1999 : 136–137]). In particular, $a_n = b_n - c_n$, where b_n (respectively c_n) denotes the total number of pairs 00, 11 (respectively, 01, 10) in the binary expansion of n. Noting that $2^k \leq n < 2^{k+1}$ if and only if the binary expansion of n has exactly $k + 1$ digits, and that the first digit is always 1, we have $a_n = 0$ if and only if $b_n = c_n$. That is we have exactly $\frac{k}{2}$ changes of digits (10 or 01) between two consecutive digits.

Show that the equation

$$x^2 + y^2 + z^2 = 3xyz$$

has infinitely many integer solutions with $x > 0$, $y > 0$ and $z > 0$.

Solution

Suppose $(1, a, b)$ is a solution with $a \leq b$. Then $a^2 + b^2 + 1 = 3ab$ and hence, $1 + b^2 + (3b - a)^2 = (1 + b^2 + a^2) + 9b^2 - 6ab = 9b^2 - 3ab = 3(1)(b)(3b - a)$. Since $3b - a > 0$, we see that $(1, b, 3b - a)$ is also a solution which is different from $(1, a, b)$ since $3b - a > b$. By iterating this, starting with the obvious solution $(1, 1, 1)$, we then obtain an infinite sequence of solutions with $x = 1$ in all of them: $(1, 1, 1)$, $(1, 1, 2)$, $(1, 2, 5)$, $(1, 5, 13)$, $(1, 13, 34)$, ...

For positive a and b and natural n prove

$$\frac{1}{a+b} + \frac{1}{a+2b} + \cdots + \frac{1}{a+nb} < \frac{n}{\sqrt{a(a+nb)}} \, .$$

Solution

By the Cauchy–Schwartz Inequality we get

$$\left(\frac{1}{a+b} + \frac{1}{a+2b} + \cdots + \frac{1}{a+nb} \right)^2$$

$$\leq n \left[\frac{1}{(a+b)^2} + \frac{1}{(a+2b)^2} + \cdots + \frac{1}{(a+nb)^2} \right]$$

$$< n \left[\frac{1}{a(a+b)} + \frac{1}{(a+b)(a+2b)} + \cdots + \frac{1}{[a+(n-1)b](a+nb)} \right]$$

$$= \frac{n}{b} \left[\frac{1}{a} - \frac{1}{a+b} + \frac{1}{a+b} - \frac{1}{a+2b} + \cdots + \frac{1}{a+(n-1)b} - \frac{1}{a+nb} \right]$$

$$= \frac{n}{b} \frac{nb}{a(a+nb)} = \frac{n^2}{a(a+nb)} \, .$$

Thus, $\dfrac{1}{a+b} + \dfrac{1}{a+2b} + \cdots + \dfrac{1}{a+nb} < \dfrac{n}{\sqrt{a(a+nb)}} \, .$

Solve the system of equations in $\mathbb{R}^3$:

$$8(x^3 + y^3 + z^3) = 73,$$
$$2(x^2 + y^2 + z^2) = 3(xy + yz + zx),$$
$$xyz = 1.$$

Solution

Let $u = x + y + z$, $v = xy + yz + zx$, $w = xyz$ and transform the given system into an equivalent system in u, v and w.

We have

$$x^2 + y^2 + z^2 = (x + y + z)^2 - 2(xy + yz + zx) = u^2 - 2v$$

and

$$x^3 + y^3 + z^3 = (x + y + z)^3 - 3(x + y + z)(xy + yz + zx) + 3xyz$$
$$= u^3 - 3uv + 3w.$$

Hence, the given system is equivalent to

$$8(u^3 - 3uv + 3w) = 73,$$
$$2(u^2 - 2v) = 3v,$$
$$w = 1,$$

whose solution is $u = \frac{7}{2}$, $v = \frac{7}{2}$, $w = 1$.

Since

$$(\zeta - x)(\zeta - y)(\zeta - z) = \zeta^3 - u\zeta^2 + v\zeta - w,$$

we see that the roots of

$$\zeta^3 - \frac{7}{2}\zeta^2 + \frac{7}{2}\zeta - 1 = 0 \qquad (1)$$

constitute a solution of the original system, and, since the equations are symmetrical, any one of the six permutations of these roots is also a solution.

The cubic (1) evidently has the root $\zeta = 1$; and the other two are easily found to be $\frac{1}{2}$ and 2.

Hence, we have the following six solutions (x, y, z):

$$\left(\frac{1}{2}, 1, 2\right), \left(\frac{1}{2}, 2, 1\right), \left(1, \frac{1}{2}, 2\right), \left(1, 2, \frac{1}{2}\right), \left(2, \frac{1}{2}, 1\right), \left(2, 1, \frac{1}{2}\right).$$

Let n and r be positive integers such that $n \geq 2$ and $r \not\equiv 0$ (mod n), and let g be the greatest common divisor of n and r. Prove that

$$\sum_{i=1}^{n-1} \left\langle \frac{ri}{n} \right\rangle = \frac{1}{2}(n - g),$$

where $\langle x \rangle = x - \lfloor x \rfloor$ is the fractional part of x.

Solution

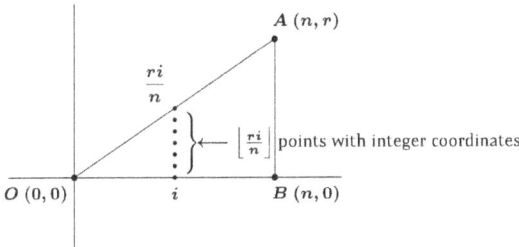

By Pick's Theorem, if P is a polygon and its vertices have integer coordinates, then we can find its area E by the formula $E = k - 1 + \frac{\ell}{2}$ where k is the number of points inside P with integer coordinates and ℓ is the number of points with integer coordinates on the sides of P.

Let $O(0,0)$, $A(n,r)$, $B(n,0)$. Then $OA : y = \frac{r}{n}x$. Since $g = (n, r)$, there are $n', r' \in \mathbb{N}$ such that $n = n'g$, $r = r'g$ and $(n', r') = 1$. If $i \in \{0, 1, \ldots, n\}$ then

$$\frac{ri}{n} \in \mathbb{N} \iff \frac{r'i}{n'} \in \mathbb{N} \iff n' \,|\, r'i \overset{(n',r')=1}{\iff} n' \,|\, i \iff i = n't, \quad t \in \mathbb{N}.$$

We have $0 \leq i \leq n \iff 0 \leq n't \leq n'g \iff 0 \leq t \leq g$.

Finally, the point $(x, \frac{rx}{n})$ on segment OA has integer coordinates if and only if $x \in \{0, 1, \ldots, n\}$ and $x = n't$, $0 \leq t \leq g$, $t \in \mathbb{N}$.

Thus, on OA there are $g + 1$ points with integer coordinates, on AB there are r more, and on OB $n-1$ more. Now $\ell = g+1+r+n-1 = g+r+n$.

Also, $k = \sum_{i=1}^{n-1} \lfloor \frac{ri}{n} \rfloor - g + 1$ since on line $x = i$ there are $\lfloor \frac{ri}{n} \rfloor$ points with integer coordinates inside triangle OAB. We exclude the $g - 1$ points which are on OA between O, A and have integer coordinates.

Now Pick's formula gives:

$$\frac{1}{2}nr = \sum_{i=1}^{n-1} \left\lfloor \frac{ri}{n} \right\rfloor - g + 1 - 1 + \frac{g+n+r}{2} \implies \sum_{i=1}^{n-1} \left\lfloor \frac{ri}{n} \right\rfloor = \frac{1}{2}nr + \frac{g-n-r}{2}.$$

Thus,

$$\begin{aligned}
\sum_{i=1}^{n-1} \left\langle \frac{ri}{n} \right\rangle &= \sum_{i=1}^{n-1} \left(\frac{ri}{n} - \left\lfloor \frac{ri}{n} \right\rfloor \right) \\
&= \frac{r}{n} \sum_{i=1}^{n-1} i - \sum_{i=1}^{n-1} \left\lfloor \frac{ri}{n} \right\rfloor = \frac{r}{n} \frac{(n-1)n}{2} - \frac{1}{2}nr + \frac{r+n-g}{2} \\
&= \frac{1}{2}nr - \frac{1}{2}r - \frac{1}{2}nr + \frac{r+n-g}{2} = \frac{1}{2}(n-g).
\end{aligned}$$

Let k and n be integers such that $1 \leq k \leq n$, and assume that a_1, $a_2, \ldots, a_k$ satisfy

$$a_1 + a_2 + \cdots + a_k = n,$$

$$a_1^2 + a_2^2 + \cdots + a_k^2 = n,$$

$$\ldots\ldots\ldots\ldots\ldots\ldots$$

$$a_1^k + a_2^k + \cdots + a_k^k = n.$$

Prove that

$$(x + a_1)(x + a_2) \cdots (x + a_k) = x^k + \binom{n}{1} x^{k-1} + \binom{n}{2} x^{k-2} + \cdots + \binom{n}{k}.$$

Solution

Let T_r denote the elementary symmetric sum of the products of the a_i's taken r at a time and let $S_r = a_1^r + a_2^r + \cdots + a_k^r$ for $r = 1, 2, \ldots, k$. It is known [1] that $T_r = \left(\frac{1}{r!}\right)$ times an $r \times r$ determinant whose successive rows are

$$
\begin{array}{ccccc}
S_1, & 1, & 0, & \ldots, & 0, \\
S_2, & S_1, & 2, & 0, \ldots, & 0, \\
\vdots & & & & \vdots \\
S_r, & S_{r-1}, & S_{r-2}, & \ldots, & S_1.
\end{array}
$$

Our proof is now by induction. Assume that $T_r = \binom{n}{r}$ for $r = 1, 2, \ldots, m$. In the determinant for T_{m+1}, subtract the mth row from the $(m+1)^{\text{st}}$ row giving the new $(m+1)^{\text{st}}$ row as $0, 0, \ldots, 0, n - m$, since all the S_j's equal n. Hence,

$$T_{m+1} = \frac{m!}{(m+1)!\,(n-m)} \binom{n}{m} = \binom{n}{m+1}$$

and hence this is valid for all m.

Solve, in integers, the equation

$$(x^2 - y^2)^2 = 1 + 16y.$$

Solution

We will prove that the solutions of

$$(x^2 - y^2)^2 = 1 + 16y \qquad (1)$$

are $(-4, 5)$, $(4, 5)$, $(-1, 0)$, $(1, 0)$, $(-4, 3)$, $(4, 3)$.

Let x, y be two integers satisfying (1). We note that:

• $y \geq 0$ unless $1 + 16y < 0$, for then $1 + 16y$ would not be a square.

• (x, y) is solution of (1). Then, with no loss of generality, we may suppose that $x \geq 0$.

Case 1: if $x \geq y$.

Then $x = y + a$ with $a \in \mathbb{N}$. Equation (1) can be written as:

$$4a^2 y^2 + 4(a^3 - 4)y + a^4 - 1 = 0.$$

Thus, y is a solution of the quadratic

$$4a^2 X^2 + 4(a^3 - 4)X + a^4 - 1 = 0.$$

Since $\Delta = 16(-8a^3 + a^2 + 16)$, we must have $f(a) = -8a^3 + a^2 + 16 \geq 0$. It is easy to see that f is decreasing for $a \in [1, +\infty)$, and that $f(2) = -44 < 0$. Then $f(a) < 0$ for $a \geq 2$.

It follows that $a \in \{0, 1\}$.

For $a = 0$, we have $x = y$, and (1) leads to $1 + 16y = 0$, which is impossible since y is an integer.

For $a = 1$, then $x = y + 1$ and (1) leads to $4y^2 - 12y = 0$. Thus, $y = 0$ or $y = 3$.

Conversely, it is easy to verify that $(1, 0)$ and $(4, 3)$ are solutions of (1) (then, so are $(-1, 0)$ and $(-4, 3)$).

Case 2: if $0 \le x < y$.

Then $0 \le x \le y - 1$ and $y \ge 1$.

It follows that

$$x^2 - y^2 \le 1 - 2y < 0$$

and then

$$1 + 16y = (x^2 + y^2)^2 \ge (1 - 2y)^2 = 1 + 4y^2 - 4y\,.$$

This leads to:

$$4y^2 - 20y \le 0\,.$$

That is, $y \in \{1, 2, 3, 4, 5\}$.

For $y \in \{1, 2, 4\}$, note that $1 + 16y$ is not a square and then (1) does not hold.

For $y = 3$, (1) leads to $(x^2 - 9)^2 = 49$ with $x^2 - 9 < 0$. Then $x^2 - 9 = -7$, which does not have any integer solution.

For $y = 5$, (1) leads to $(x^2 - 25)^2 = 81$ with $x^2 - 25 < 0$. Then $x^2 - 25 = -9$ and $x = 4$.

Conversely, it is easy to verify that $(4, 5)$ is a solution of (1) (and then so is $(-4, 5)$), which completes the proof.

Given all possible quadratic trinomials of the type $x^2 + px + q$, with integer coefficients p and q, $1 \leq p \leq 1997$, $1 \leq q \leq 1997$. Consider the sets of the trinomials:

(a) having integer zeros,

(b) not having real zeros.

Which of those sets is larger?

Solution

We will prove that there are more trinomials not having real zeros than trinomials having integer zeros.

Let $G = \{1, \ldots, 1997\}$. Note that such a trinomial has integer zeros if and only if (p, q) belongs to

$$F_1 = \{(p, q) \in G^2 : p^2 - 4q \text{ is a perfect square,}\}$$

and that a trinomial does not have real zeros if and only if (p, q) belongs to

$$F_2 = \{(p, q) \in G^2 : p^2 - 4q < 0\}$$

Thus, we will prove that $\text{Card}(F_2) > \text{Card}(F_1)$.

Let $(p, q) \in F_1$. Then there exists a non-negative integer a such that

$$p^2 - 4q = a^2.$$

Thus, $a = \sqrt{p^2 - 4q}$, a and p have the same parity, and (since $q \geq 1$) we have $a < p$. It follows that $a \leq p - 2$. Thus, $p - a - 1 \in \mathbb{N}^*$. Moreover,

$$
\begin{aligned}
(p - a - 1)^2 - 4q &= p^2 - 4q + a^2 + 1 - 2ap - 2p + 2a \\
&= 2a^2 + 1 - 2ap + 2a - 2p \\
&= 2a(a - p) + 1 - 2(p - a),
\end{aligned}
$$

where $2a(a - p) \leq 0$ and $1 - 2(p - a) < 0$. Thus, $(p - a - 1)^2 - 4q < 0$.

Since $p - a - 1 \in G$, we deduce that $(p - 1 - \sqrt{p^2 - 4q}, q)$ belongs to F_2.

Let

$$
\begin{aligned}
f : F_1 &\longrightarrow F_2 \\
(p, q) &\mapsto (p - 1 - \sqrt{p^2 - 4q}, q).
\end{aligned}
$$

If $(p, q), (p', q') \in F_1$ such that $f(p, q) = f(p', q')$, then $q = q'$ and

$$p - 1 - \sqrt{p^2 - 4q} = p' - 1 - \sqrt{(p')^2 - 4q}.$$

Thus,

$$p - p' = \sqrt{p^2 - 4q} - \sqrt{(p')^2 - 4q}.$$

Suppose that $p \neq p'$. Then $\sqrt{p^2 - 4q} + \sqrt{(p')^2 - 4q} > 0$ and we have

$$p - p' = \frac{(p - p')(p + p')}{\sqrt{p^2 - 4q} + \sqrt{(p')^2 - 4q}}.$$

Thus,

$$p + p' = \sqrt{p^2 - 4q} + \sqrt{(p')^2 - 4q}. \qquad (1)$$

Since $q \geq 1$, we have

$$\sqrt{p^2 - 4q} < \sqrt{p^2} = p$$

and

$$\sqrt{p'^2 - 4q} < \sqrt{(p')^2} = p'.$$

Thus, (1) is not satisfied. A contradiction.

Therefore, $p = p'$, and f is injective.

It follows that (since F_1 and F_2 are finite):

$$\mathrm{Card}(F_2) \geq \mathrm{Card}(F_1),$$

and it now suffices to prove that f is not surjective.

Suppose, for a contradiction, that there exists $(p, q) \in F_1$ such that $f(p, q) = (2, 3)$. Then $q = 3$ and

$$p - 1 - \sqrt{p^2 - 12} = 2.$$

Therefore,

$$(p - 3)^2 = p^2 - 12,$$

and thus, $p^2 - 6p + 9 = p^2 - 12$ giving $6p = 21$, which has no integral solution for p. It follows that there is no $(p, q) \in F_1$ such that $f(p, q) = (2, 3)$. Since $(2, 3) \in F_2$, we deduce that f is not surjective. Then $\mathrm{Card}(F_2) > \mathrm{Card}(F_1)$ and the proof is complete.

(a) Prove that for all real numbers p and q the inequality $p^2 + q^2 + 1 > p(q + 1)$ holds.

(b) Determine the greatest real number b such that for all real numbers p and q the inequality $p^2 + q^2 + 1 > bp(q + 1)$ holds.

(c) Determine the greatest real number c such that for all integers p and q the inequality $p^2 + q^2 + 1 > cp(q + 1)$ holds.

Solution

(a)–(b) It is easy to see that we can suppose $p, q \geq 0$. We have

$$p^2 + q^2 + 1 \geq p^2 + 2\left(\frac{q+1}{2}\right)^2 \geq 2\sqrt{\frac{p^2(q+1)^2}{2}} = \sqrt{2}\,p(q+1);$$

equality holds for $p = \sqrt{2}$, $q = 1$. Hence $b = \sqrt{2}$.

(c) If $p = q = 1$, we have $1 + 1 + 1 = \frac{3}{2}1(1+1)$, whence $\frac{3}{2}$ is an upper bound for the maximum. In fact, it is the maximum, since:

- if $q = 0$, then $p^2 + 1 > \frac{3}{2}p$, true.
- if $q = 1$, then $p^2 + 2 \geq 3p$; that is, $(p-1)(p-2) \geq 0$, true.
- if $q = 2$, then $p^2 + 5 \geq \frac{9}{2}p$; that is, $(p-2)(p-\frac{5}{2}) \geq 0$, true.
- if $q \geq 3$, then $p(3-q) - 2 < 0 \leq 2(p-q)^2$; hence

$$p^2 + q^2 + 1 > \frac{3}{2}p(q+1).$$

Find all real solutions of

$$\sqrt[4]{13 + x} + \sqrt[4]{4 - x} = 3.$$

Solution

Let $u = \sqrt[4]{13 + x}$ and $v = \sqrt[4]{4 - x}$. Then $u + v = 3$ and $u^4 + v^4 = 17$. Thus,

$$17 + 2u^2v^2 = (u^2 + v^2)^2 = ((u + v)^2 - 2uv)^2 = (9 - 2uv)^2.$$

Simplifying, we get $u^2v^2 - 18uv + 32 = 0$, whence $(uv - 2)(uv - 16) = 0$.

If $uv = 2$, then $(13 + x)(4 - x) = 16$ yields $x^2 + 9x - 36 = 0$ or $(x - 3)(x + 12) = 0$. Hence $x = 3$ or -12.

If $uv = 16$, then $(13 + x)(4 - x) = 16^4$ yields $x^2 + 9x + 65484 = 0$ which clearly has no real solutions.

Since it is easy to see that both $x = 3$ and $x = -12$ satisfy the given equation, we conclude that the only real solutions are $x = 3$ and $x = -12$.

Find

(a) all quadruples of positive integers (a, k, l, m) for which the equality $a^k = a^l + a^m$ holds;

(b) all 5-tuples of positive integers (a, k, l, m, n) for which the equality $a^k = a^l + a^m + a^n$ holds.

Solution

(a) Obviously $a \geq 2$ and $k \geq l, m$. If $l \neq m$, for example, $l > m$, then $a^{k-m} = a^{l-m} + 1$ with $k - m, l - m \geq 1$. Therefore, a divides $a^{k-m} - a^{l-m} = 1$, a contradiction.

Thus, $l = m$ and $a^k = 2 \cdot a^m \iff a^{k-m} = 2 \iff a = 2, k - m = 1$. All quadruples are $(2, m + 1, m, m)$, $m = 1, 2, 3, \ldots$.

(b) Obviously $a \geq 2$, $k \geq l, m, n$. Without loss of generality, we suppose that $l \geq m \geq n$. As above, we see that $m = n$, so the equation becomes $a^{k-m} = a^{l-m} + 2$.

- If $l = m$, then $a^{k-m} = 3$, then $a = 3, k - m = 1$.
- If $l > m$, then $a|2$. Thus, $a = 2$ and $2^{k-m} = 2^{l-m} + 2$. By (a), $k - m = 2$, $l - m = 1$.

All 5-tuples are $(3, m + 1, m, m, m)$, $(2, m + 2, m + 1, m, m)$, where $m = 1, 2, 3, \ldots$, with appropriate permutations.

Prove that, for any real numbers x and y, the following inequality holds:

$$x^2 + y^2 + 1 > x\sqrt{y^2 + 1} + y\sqrt{x^2 + 1}.$$

Solution

We have $\left(x - \sqrt{y^2 + 1}\right)^2 \geq 0$, $\left(y - \sqrt{x^2 + 1}\right)^2 \geq 0$.

Hence,

$$\left(x - \sqrt{y^2 + 1}\right)^2 + \left(y - \sqrt{x^2 + 1}\right)^2 \geq 0.$$

That is,

$$x^2 + y^2 + 1 \geq x\sqrt{y^2 + 1} + y\sqrt{x^2 + 1}.$$

The equality holds if and only if $x^2 = y^2 + 1$ and $y^2 = x^2 + 1$, a system which has no solutions. Consequently,

$$x^2 + y^2 + 1 > x\sqrt{y^2 + 1} + y\sqrt{x^2 + 1}.$$

For positive integers m, n denote $T(m,n) = \gcd\left(m, \frac{n}{\gcd(m,n)}\right)$.

(a) Prove that there exist infinitely many pairs of integers (m,n) such that $T(m,n) > 1$ and $T(n,m) > 1$.

(b) Does there exist a pair of integers (m,n) such that $T(m,n) = T(n,m) > 1$?

Solution

In what follows we simply denote $\gcd(a,b)$ by (a,b).

(a) Let $(m,n) = d$, $m = dm_1$, $n = dn_1$, $(m_1, n_1) = 1$. Now we have

$$T(m,n) = (dm_1, n_1) > 1 \quad \text{if and only if } (d, n_1) > 1.$$

Similarly, we obtain

$$T(n,m) = (dn_1, m_1) > 1 \quad \text{if and only if } (d, m_1) > 1.$$

It is clear that there are infinitely many pairs (m,n) which satisfy these two conditions: for example, if we set $d = 2^p 3^q$, $m_1 = 2$ and $n_1 = 3$, we obtain infinitely many such pairs, since there are infinitely many pairs (p,q) with natural numbers p, q.

(b) If we have the required relation, then using the notation from (a) we must have

$$(dn_1, m_1) = (dm_1, n_1) > 1 \quad \Longleftrightarrow \quad (d, m_1) = (d, n_1) = d_1.$$

It follows that $d_1 = 1$, since otherwise we would have $(n_1, m_1) > 1$. Hence, there are no integers which satisfy the conditions.

Solve the system in real numbers

$$\begin{cases} x_1 + x_2 + \cdots + x_{1997} & = & 1997 \\ x_1^4 + x_2^4 + \cdots + x_{1997}^4 & = & x_1^3 + x_2^3 + \cdots + x_{1997}^3 . \end{cases}$$

Solution

More generally, let n be a positive integer. We will prove that the unique solution in real numbers of the system

$$\begin{cases} x_1 + x_2 + \cdots + x_n & = & n \\ x_1^4 + x_2^4 + \cdots + x_n^4 & = & x_1^3 + x_2^3 + \cdots + x_n^3 \end{cases}$$

is $(1, 1, \ldots, 1)$. It is easy to see that $(1, 1, \ldots, 1)$ is a solution of the system. Conversely, if $x_1, x_2, \ldots, x_n$ are real numbers satisfying the system, then

$$\begin{aligned} 0 & = \sum_{i=1}^{n} x_i^4 - \sum_{i=1}^{n} x_i^3 - \sum_{i=1}^{n} x_i + n = \sum_{i=1}^{n} (x_i^4 - x_i^3 - x_i + 1) \\ & = \sum_{i=1}^{n} (x_i - 1)^2 (x_i^2 + x_i + 1) = \sum_{i=1}^{n} (x_i - 1)^2 \left(\left(x_i + \frac{1}{2} \right)^2 + \frac{3}{4} \right) . \end{aligned}$$

Since, for each i, we have $(x_i - 1)^2 \left((x_i + \frac{1}{2})^2 + \frac{3}{4} \right) \geq 0$, the equality occurs only if $x_i = 1$ for $i = 1, 2, \ldots, n$.

By Chebyshev's inequality, if $a_1 \leq \cdots \leq a_n$ and $b_1 \leq \cdots \leq b_n$, then

$$n(a_1 b_1 + a_2 b_2 + \cdots + a_n b_n) \geq (a_1 + a_2 + \cdots + a_n)(b_1 + b_2 + \cdots + b_n) .$$

Equality occurs if and only if at least one sequence is constant [2001 : 514].

Without loss of generality, we assume that $x_1 \leq x_2 \leq \cdots \leq x_{1997}$. Then $x_1^3 \leq x_2^3 \leq \cdots \leq x_{1997}^3$. Therefore,

$$1997(x_1^4 + x_2^4 + \cdots + x_{1997}^4) \geq (x_1 + x_2 + \cdots + x_{1997})(x_1^3 + x_2^3 + \cdots + x_{1997}^3) .$$

Since $x_1 + x_2 + \cdots + x_{1997} = 1997$ and

$$x_1^4 + x_2^4 + \cdots + x_{1997}^4 = x_1^3 + x_2^3 + \cdots + x_{1997}^3 ,$$

the above inequality becomes equality. Thus,

$$x_1 = x_2 = \cdots = x_{1997} \quad \text{or} \quad x_1^3 = x_2^3 = \cdots = x_{1997}^3 .$$

In either case, $x_1 = x_2 = \cdots = x_{1997} = 1$.

It is known that the equation $ax^3 + bx^2 + cx + d = 0$
with respect to x has three distinct real roots. How many roots does the
equation $4(ax^3 + bx^2 + cx + d)(3ax + b) = (3ax^2 + 2bx + c)^2$ have?

Solution

Let $p(x) = ax^3 + bx^2 + cx + d$, and let

$$f(x) = 4(ax^3 + bx^2 + cx + d)(3ax + b) - (3ax^2 + 2bx + c)^2 \,.$$

We are told that the equation $p(x) = 0$ has three distinct real roots, and we
are asked about the roots of the equation $f(x) = 0$. Clearly, the problem
is interested in real roots only. Note that $f(x) = 2p(x)p''(x) - [p'(x)]^2$.
Hence, $f'(x) = 2p(x)p'''(x) = 12ap(x)$ and $f''(x) = 12ap'(x)$.

Let the roots of $p(x)$ be r_1, r_2, and r_3, where $r_1 < r_2 < r_3$. Since
these roots are distinct, $p'(r_i) \neq 0$ (for $i = 1, 2, 3$). Since there are three
roots, the degree of $p(x)$ cannot be less than three. Therefore, $a \neq 0$. We
can assume $a > 0$. (Otherwise we replace $p(x)$ by $-p(x)$, with no effect
on $f(x)$.) Then $\lim_{x \to \infty} p(x) = \infty$ and $\lim_{x \to -\infty} p(x) = -\infty$. Hence
$p'(r_1) > 0$, $p'(r_2) < 0$ and $p'(r_3) > 0$.

For each i, $f'(r_i) = 12ap(r_i) = 0$, and there are no other points where
$f'(x) = 0$. Since $f''(x)$ has the same sign as $p'(x)$, we have $f''(r_1) > 0$,
$f''(r_2) < 0$ and $f''(r_3) > 0$. Thus, f has local minima at r_1 and r_3, a
local maximum at r_2, and no other local maxima or minima. For each i,
$f(r_i) = -[p'(r_i)]^2 < 0$. Therefore, $f(x) < 0$ on an interval containing
r_1, r_2 and r_3. Furthermore, $\lim_{x \to \pm\infty} f(x) = \infty$ (since the highest-degree
term in $f(x)$ is $3a^2x^4$). All of this implies that the equation $f(x) = 0$ has
exactly two real roots, one of which is less than r_1 (where $f(x)$ changes from
positive to negative) and the other greater than r_3 (where $f(x)$ changes from
negative to positive).

Find (with proof) all pairs of integers (x, y) satisfying the equation

$$1 + 1996x + 1998y = xy.$$

Solution

We find all pairs of integers (x, y) satisfying the more general equation

$$1 + (p-1)x + (p+1)y = xy,$$

where $p > 1$ is a prime number. This equation is equivalent to

$$px + py = xy + x - y - 1,$$

which can be rewritten as

$$p((x-1) + (y+1)) = (x-1)(y+1). \tag{1}$$

We observe that $(x, y) = (1, -1)$ is a solution and that no other candidates for (x, y) with $x = 1$ or $y = -1$ can satisfy (1).

Now suppose that $x \neq 1$ and $y \neq -1$, and denote by d the greatest common divisor of $x - 1$ and $y + 1$. We have

$$x - 1 = du, \quad y + 1 = dv, \tag{2}$$

where u and v are relatively prime integers. Substituting these expressions for $x - 1$ and $y + 1$ into (1) and dividing both sides by d gives

$$p(u + v) = duv. \tag{3}$$

Hence, uv divides the product $p(u + v)$ and is relatively prime to $u + v$. By the Fundamental Theorem of Arithmetic, uv is a divisor of p. Thus,

$$uv \in \{1, -1, p, -p\}.$$

Since $p > 0$ and $d > 0$, it follows from (3) that $u + v$ and uv agree in sign. This leads to the following possibilities:

- $u = v = 1$. Then, by (3), $d = 2p$. Substituting these values into (2), we find that
$$x = 2p + 1, \quad y = 2p - 1.$$

- $u = 1$, $v = p$. This yields $d = p + 1$ and
$$x = p + 2, \quad y = p(p + 1) - 1.$$

- $u = 1$, $v = -p$. This yields $d = p - 1$ and
$$x = p, \quad y = -p(p - 1) - 1.$$

- $u = p$, $v = 1$. This yields $d = p + 1$ and
$$x = p(p + 1) + 1, \quad y = p.$$

- $u = -p$, $v = 1$. This yields $d = p - 1$ and

$$x \;=\; 1 - p(p-1), \quad y \;=\; p - 2.$$

We conclude that the set of solutions for (x, y) is

$$\{(1, -1), \quad (2p+1, 2p-1), \quad (p+2, p(p+1) - 1), \quad (p, -p(p-1) - 1),$$
$$(p(p+1) + 1, p), \quad (1 - p(p-1), p - 2)\}.$$

The given problem is the special case when $p = 1997$.

Let a, b, c be non-negative real numbers such that $a + b + c \geq abc$.
Prove that $a^2 + b^2 + c^2 \geq abc$.

Solution

More generally, let $n \geq 2$ be an integer and $1 \leq p \leq n$. We claim that
if $a_1, a_2, \ldots, a_n$ are non-negative real numbers such that $\sum a_i \geq \prod a_i$, then

$$\sum a_i^p \geq n^{(p-1)/(n-1)} \prod a_i .$$

(Here and below, all sums and products are extended over $i = 1, 2, \ldots, n$.)

Proof. We consider two cases:

Case 1. $\prod a_i \leq n^{n/(n-1)}$.
From the AM-GM Inequality, we have

$$\sum a_i^p \geq n \left(\prod a_i^p \right)^{1/n} = n \left(\prod a_i \right)^{(p-n)/n} \prod a_i$$
$$\geq n^{(p-1)/(n-1)} \prod a_i ,$$

because $p \leq n$.

Case 2. $\prod a_i > n^{n/(n-1)}$.
Using the Power-Mean Inequality and the condition $\sum a_i \geq \prod a_i$, we
obtain

$$\left(\frac{1}{n} \sum a_i^p \right)^{1/p} \geq \frac{1}{n} \sum a_i \geq \frac{1}{n} \prod a_i ,$$

which implies

$$\sum a_i^p \geq n^{1-p} \left(\prod a_i \right)^p = n^{1-p} \left(\prod a_i \right)^{p-1} \prod a_i$$
$$\geq n^{(p-1)/(n-1)} \prod a_i .$$

This completes the proof of the claim.

Taking $n = 3$ and $p = 2$, and renaming a_1, a_2, a_3 by a, b, c, we see that
under the conditions given in the proposal, there holds the better inequality
$a^2 + b^2 + c^2 \geq \sqrt{3}\, abc$. This inequality is stronger than the one proposed.

We also give Klamkin's solution and remarks.
We need only consider two cases.

(1) If a, b, c are all ≥ 1, then clearly $a^2 + b^2 + c^2 \geq a + b + c \geq abc$.

(2) If at least one of a, b, $c \leq 1$ (say $c \leq 1$), then $a^2 + b^2 \geq ab \geq abc$.

We now conclude the proof by allowing negative numbers. If just one of a, b, c is negative or if all three are negative, then $abc \leq 0$, in which case the result is immediate. Thus, we may assume that only two of them, say b and c, are negative. Then, letting $x = -b$ and $y = -c$, we have to show that when $a \geq x + y + axy$, we can conclude that $a^2 + x^2 + y^2 \geq axy$.

The assumed inequality implies that $xy \leq 1$. If $a \geq 1$, then $a^2 \geq axy$, which implies the desired result. If $a \leq 1$, then $x \leq 1$ and $y \leq 1$. In this case $a > x + y \geq xy$, since the latter inequality can be rewritten as $1 \geq (1 - x)(1 - y)$. Thus, we still have $a^2 \geq axy$, with the same conclusion as before.

Let S be the set of all odd integers greater than one. For each $x \in S$, denote by $\delta(x)$ the unique integer satisfying the inequality

$$2^{\delta(x)} < x < 2^{\delta(x)+1}.$$

For $a, b \in S$, define

$$a * b = 2^{\delta(a)-1}(b - 3) + a.$$

[For example, to calculate $5 * 7$ note that $2^2 < 5 < 2^3$, so that $\delta(5) = 2$, and hence, $5 * 7 = 2^{2-1}(7 - 3) + 5 = 13$. Also $2^2 < 7 < 2^3$, so that $\delta(7) = 2$ and $7 * 5 = 2^{2-1}(5 - 3) + 7 = 11$].

Prove that if $a, b, c \in S$, then

(a) $a * b \in S$ and

(b) $(a * b) * c = a * (b * c)$.

Solution

(a) Since $b-3$ is even, $a*b$ is clearly odd. Also, since $2^{\delta(a)-1}(b-3) \geq 0$, we have $a * b \geq a > 1$. Hence, $a * b \in S$.

(b) Note first that

$$
\begin{aligned}
(a * b) * c &= 2^{\delta(a*b)-1}(c - 3) + (a * b) \\
&= 2^{\delta(a*b)-1}(c - 3) + 2^{\delta(a)-1}(b - 3) + a \qquad (1)
\end{aligned}
$$

and

$$
\begin{aligned}
a * (b * c) &= a * (2^{\delta(b)-1}(c - 3) + b) \\
&= 2^{\delta(a)+\delta(b)-2}(c - 3) + 2^{\delta(a)-1}(b - 3) + a . \qquad (2)
\end{aligned}
$$

By (1) and (2) it clearly suffices to show that

$$\delta(a * b) = \delta(a) + \delta(b) - 1 . \qquad (3)$$

By definition, $2^{\delta(a)} < a < 2^{\delta(a)+1}$ and $2^{\delta(b)} < b < 2^{\delta(b)+1}$.

Using the inequalities $a < 2^{\delta(a)+1}$ and $b < 2^{\delta(b)+1}$, we have

$$
\begin{aligned}
2^{\delta(a)-1}(b - 3) + a &< 2^{\delta(a)-1}(b - 3) + 2^{\delta(a)+1} \\
&= 2^{\delta(a)-1}(b - 3 + 4) \\
&= 2^{\delta(a)-1}(b + 1) \leq 2^{\delta(a)-1} \cdot 2^{\delta(b)+1} ;
\end{aligned}
$$

Is there an integer N such that

$$(\sqrt{1997} - \sqrt{1996})^{1998} = \sqrt{N} - \sqrt{N-1}?$$

Solution

One such integer is

$$N = \left(\frac{(\sqrt{1997} + \sqrt{1996})^{1998} + (\sqrt{1997} - \sqrt{1996})^{1998}}{2} \right)^2.$$

In the following, we shall establish a more general result.

Theorem. Let a, $b \in \mathbb{R}$, with $0 \le b \le a$. Then for all $n \in \mathbb{N}$, we have $(a - b)^n = \sqrt{k^2} - \sqrt{k^2 - (a^2 - b^2)^n}$, where $k = \frac{1}{2}((a + b)^n + (a - b)^n)$.

Proof. Since

$$k^2 - \left(\frac{(a + b)^n - (a - b)^n}{2} \right)^2 = (a^2 - b^2)^n,$$

we get

$$\sqrt{k^2} - \sqrt{k^2 - (a^2 - b^2)^n} = k - \frac{(a + b)^n - (a - b)^n}{2} = (a - b)^n.$$

Corollary 1. Let d, m, $n \in \mathbb{N}$, with $d \le m$. Then

$$(\sqrt{m} - \sqrt{m - d})^n = \sqrt{k^2} - \sqrt{k^2 - d^n},$$

where $k = \frac{1}{2}((\sqrt{m} + \sqrt{m - d})^n + (\sqrt{m} - \sqrt{m - d})^n)$.

Proof. In the theorem, let $a = \sqrt{m}$ and $b = \sqrt{m - d}$.

Corollary 2. Let m, $n \in \mathbb{N}$. Then

$$(\sqrt{m} - \sqrt{m - 1})^n = \sqrt{k^2} - \sqrt{k^2 - 1},$$

where $k = \frac{1}{2}((\sqrt{m} + \sqrt{m - 1})^n + (\sqrt{m} - \sqrt{m - 1})^n)$.

The given problem is the special case of Corollary 2 when $m = 1997$, $n = 1998$, and $N = k^2$. Note that $k \in \mathbb{N}$, since n is even and

$$k = \sum_{i=0}^{\lfloor n/2 \rfloor} \binom{n}{2i} (\sqrt{m})^{n-2i} \left(\sqrt{m - 1} \right)^{2i}.$$

Find all real numbers α with the following property: for any positive integer n there exists an integer m such that

$$\left|\alpha - \frac{m}{n}\right| < \frac{1}{3n}.$$

Solution

We will prove that the real numbers α with the specified property are the integers.

Note that if m, n are integers with $n > 0$, the condition $\left|\alpha - \frac{m}{n}\right| < \frac{1}{3n}$ may be rewritten as $\left|n\alpha - m\right| < \frac{1}{3}$, which is the same as $d(n\alpha, m) < \frac{1}{3}$, where d denotes the usual distance. Thus, the property described in the problem is equivalent to the following property P:

$$d(n\alpha, \mathbb{Z}) < \frac{1}{3}, \qquad \text{for any positive integer } n.$$

Case 1. Suppose α is irrational. Then, from a well-known theorem of Kronecker (see [1]), the set $\{n\alpha - \lfloor n\alpha \rfloor \mid n \in \mathbb{N}^*\}$ is dense in $[0, 1]$. Hence, there exists $n \in \mathbb{N}^*$ such that $d\left(n\alpha - \lfloor n\alpha \rfloor, \frac{1}{2}\right) < 0.1$. Let $a = \lfloor n\alpha \rfloor$. Then $a + 0.4 < n\alpha < a + 0.6$ which leads to

$$d(n\alpha, \mathbb{Z}) = \min\{d(n\alpha, a), d(n\alpha, a + 1)) > 0.4 > \frac{1}{3}.$$

Thus, α does not have the property P.

Case 2. Suppose α is rational. Let $\alpha = \frac{a}{b}$ where a, b are relatively prime integers and $b > 1$.

- If $b = 2k$ is even, then a is odd, and we have $k\alpha = a/2$. It follows that $d(k\alpha, \mathbb{Z}) = \frac{1}{2} > \frac{1}{3}$. Thus, α does not have the property P.

- If $b = 2k + 1$ is odd and $k > 1$, we note that since a, b are relatively prime, there exist integers u, v with $u > 0$ such that $au + bv = 1$. Then

$$u\alpha = \frac{ua}{b} = \frac{1 - bv}{b} = -v + \frac{1}{b} = -v + \frac{1}{2k + 1},$$

and hence,

$$ku\alpha = -kv + \frac{k}{2k + 1}.$$

Then $d(ku\alpha, \mathbb{Z}) = \frac{k}{2k + 1}$ or $d(ku\alpha, \mathbb{Z}) = 1 - \frac{k}{2k + 1} = \frac{k + 1}{2k + 1}$. In either case $d(ku\alpha, \mathbb{Z}) > \frac{1}{3}$ (noting that $k > 1$). Thus, α does not have the property P.

- If $b = 3$, then $\alpha = p + \frac{c}{3}$ where p is an integer and $c \in \{1, 2\}$. Then $d(\alpha, \mathbb{Z}) = \frac{1}{3}$, and α does not have the property P.

We have shown that if $\alpha \notin \mathbb{Z}$, then α does not have the property P. Moreover, if $\alpha \in \mathbb{Z}$, then obviously α has the property P (choose $m = n\alpha$). Thus, α has the property P if and only if α is an integer, as claimed.

Positive integers x, y, z satisfy the equation $2x^x + y^y = 3z^z$. Prove that they are equal.

Solution

Let x, y, z be positive integers, such that $2x^x + y^y = 3z^z$.

Suppose that $y > z$. Then, since they are integers, we have $y \geq z + 1$. Using the Binomial Theorem,

$$3z^z \ = \ 2x^x + y^y \ > \ (z+1)^{z+1} \ \geq \ z^{z+1} + (z+1)z^z \ \geq \ 3z^z.$$

This is a contradiction. Thus, $y \leq z$.

If $y < z$, then $2x^x = 3z^z - y^y > 2z^z$, and so $x > z$. As above, we then have $x \geq z + 1$ and

$$3z^z \ = \ 2x^x + y^y \ > \ 2(z+1)^{z+1} \ \geq \ 2z^{z+1} + (z+1)z^z \ > \ 3z^z.$$

This is a contradiction. Thus, $y = z$. Then $x^x = z^z$, which gives $x = z$, and we are done.

Prove that for $x \geq 2$, $y \geq 2$, $z \geq 2$

$$(y^3 + x)(z^3 + y)(x^3 + z) \geq 125xyz.$$

Solution

We show that

$$(x_1^{nr} + ax_2^{ns})(x_2^{nr} + ax_3^{ns}) \cdots (x_n^{nr} + ax_1^{ns}) \geq (k^{n(r-s)} + a)^n P^{ns}, \quad (1)$$

where $P = x_1 x_2 \cdots x_n$, $r \geq s \geq 0$, $a \geq 0$, and $x_i \geq k \geq 0$.

Since $k^n \leq P$, we have $k^{n(r-s)} \leq P^{r-s}$, and hence,

$$(k^{n(r-s)} + a)P^s \leq P^r + aP^s$$
$$\leq (x_1^{nr} + ax_2^{ns})^{\frac{1}{n}}(x_2^{nr} + ax_3^{ns})^{\frac{1}{n}} \cdots (x_n^{nr} + ax_1^{ns})^{\frac{1}{n}},$$

where the last step follows by Hölder's Inequality. Now raise both sides to the power n to obtain (1). There is equality if and only if $x_i = k$ for all i.

The given inequality corresponds to the special case $n = 3$, $r = 1$, $s = \frac{1}{3}$, $a = 1$ and $k = 2$.

Other extensions can be obtained by replacing each of the factors on the left side of (1) by a sum of more terms, and using Hölder's Inequality. For example, the first factor can be replaced by $x_1^{nr} + ax_2^{ns} + bx_3^{nt}$, with the other $n - 1$ factors given cyclically.

Find the greatest real number α such that there exists an infinite sequence of whole numbers (a_n) $(n = 1, 2, 3, \ldots)$ satisfying simultaneously the following conditions:

(i) $a_n > 1997^n$ for every $n \in \mathbb{N}^*$,

(ii) $a_n^\alpha \le U_n$ for every $n \ge 2$, where U_n is the greatest common divisor of the set of numbers $\{a_i + a_j \mid i + j = n\}$.

Solution

The greatest number α is $\frac{1}{2}$.

Let F_n be the nth Fibonacci number. Let m be an even positive integer such that $F_{2mn} > 1997^n$ for all $n \in \mathbb{N}^*$, and let $a_n = 3F_{2mn}$. The sequence $(a_n)_{n=1}^\infty$ satisfies condition (i). We will prove that it satisfies condition (ii) with $\alpha = \frac{1}{2}$.

The well-known recurrence relation $F_{n+2} = F_{n+1} + F_n$ serves to define F_n for all integers $n \ge 0$, given that $F_0 = 0$ and $F_1 = 1$. We can extend this definition by letting $F_{-n} = (-1)^{n+1}F_n$. Alternatively, we can define F_n for any integer n by the explicit formula

$$F_n = \frac{1}{\sqrt{5}}\left[\left(\frac{1+\sqrt{5}}{2}\right)^n - \left(\frac{1-\sqrt{5}}{2}\right)^n\right].$$

For any integers k and ℓ, we have $F_{k+\ell} = F_k F_{\ell+1} + F_{k-1}F_\ell$. This identity may be taken as the basis for several of the assertions below.]

For all positive integers i and j, we have

$$F_{2mi} = F_{m(i+j)}F_{m(i-j)+1} + F_{m(i+j)-1}F_{m(i-j)}.$$

Interchanging i and j and recalling that m is even, we obtain

$$F_{2mj} = F_{m(i+j)}F_{m(i-j)-1} - F_{m(i+j)-1}F_{m(i-j)}.$$

Hence,

$$F_{2mi} + F_{2mj} = F_{m(i+j)}\left(F_{m(i-j)+1} + F_{m(i-j)-1}\right).$$

If $i + j = n$, then $F_{mn} \mid (F_{2mi} + F_{2mj})$, and hence, $3F_{mn} \mid (a_i + a_j)$, which implies that $3F_{mn} \le U_n$ (where U_n is defined as in the problem statement above). Furthermore,

$$F_{2mn} = F_{mn}\left(F_{mn+1} + F_{mn-1}\right) = F_{mn}\left(2F_{mn} + F_{mn-1}\right) \le 3F_{mn}^2.$$

Thus, $a_n = 3F_{2mn} \le 9F_{mn}^2 \le U_n^2$, and finally, $a_n^{1/2} \le U_n$, as required in condition (ii).

Now, consider any $\alpha > 0$, and suppose that some sequence (a_n) satisfies the conditions (i) and (ii) for this α. We will prove that $\alpha \le \frac{1}{2}$.

Consider any ε such that $0 < \varepsilon < 2$. Suppose that there exists $N \in \mathbb{N}^*$ such that $a_{2n} < a_n^{2-\varepsilon}$ for all $n > N$. Then, for any $n > N$, we have $\frac{\log a_{2n}}{2n} < \left(\frac{2-\varepsilon}{2} \right) \left(\frac{\log a_n}{n} \right)$, and therefore, $\frac{\log a_{2^k n}}{2^k n} < \left(\frac{2-\varepsilon}{2} \right)^k \left(\frac{\log a_n}{n} \right)$ for all $k \in \mathbb{N}^*$. Then $\lim\limits_{k \to \infty} \frac{\log a_{2^k n}}{2^k n} = 0$. But this is impossible, because condition (i) implies that $\frac{\log a_n}{n} \ge \log 1997$ for all n. Consequently, there must be infinitely many values of n for which $a_{2n} \ge a_n^{2-\varepsilon}$. For any such n,

$$a_n^{(2-\varepsilon)\alpha} \le a_{2n}^\alpha \le U_{2n} \le 2a_n \,,$$

and hence, $2 \ge a_n^{(2-\varepsilon)\alpha-1} \ge 1997^{n((2-\varepsilon)\alpha-1)}$. Therefore, $(2-\varepsilon)\alpha - 1 \le 0$. Letting $\varepsilon \to 0$, we have $\alpha \le \frac{1}{2}$.

Determine all pairs of positive real numbers a, b such that for every $n \in \mathbb{N}^*$ and for every real root x_n of the equation

$$4n^2 x = \log_2(2n^2 x + 1)$$

we have

$$a^{x_n} + b^{x_n} \geq 2 + 3x_n \,.$$

Solution

We have

$$
\begin{aligned}
4n^2 x = \log_2(2n^2 x + 1) &\iff 4^{2n^2 x} = 2n^2 x + 1 \\
&\iff 2n^2 x = -\tfrac{1}{2} \text{ or } 0 \\
&\iff x = -\tfrac{1}{4n^2} \text{ or } 0 \,.
\end{aligned}
$$

Thus, the set of roots x_n is $E = \{0\} \cup \{-\frac{1}{4n^2} \mid n \in \mathbb{N}^*\}$. We claim that $a^x + b^x \geq 2 + 3x$ for all $x \in E$ if and only if $ab \leq e^3$.

Note first that

$$\lim_{x \to 0} \frac{a^x - 1}{x} = \log a \,, \qquad \lim_{x \to 0} \frac{b^x - 1}{x} = \log b \,,$$

and hence,

$$\lim_{x \to 0} \frac{a^x + b^x - 2}{x} = \log(ab) \,.$$

If $a^x + b^x \geq 2 + 3x$ for all $x \in E$, then for all n,

$$\frac{a^{-\frac{1}{4n^2}} + b^{-\frac{1}{4n^2}} - 2}{-\frac{1}{4n^2}} \leq 3 \,.$$

Letting $n \to \infty$, we find that $\log(ab) \leq 3$, and therefore, $ab \leq e^3$.

Conversely, let us suppose that $ab \leq e^3$. Using the Power–Mean Inequality, we have, for $x < 0$,

$$\left(\frac{a^x + b^x}{2} \right)^{1/x} \leq \sqrt{ab} \leq e^{3/2} \,.$$

Then

$$\frac{a^x + b^x}{2} \geq e^{3x/2} \geq 1 + \frac{3x}{2} \,,$$

and thus, $a^x + b^x \geq 2 + 3x$. This inequality is also satisfied when $x = 0$, trivially. Therefore, it is satisfied for all $x \in E$.

Let a, b, c be positive integers. Prove that the inequality

$$\frac{(b+c-a)^2}{(b+c)^2+a^2} + \frac{(c+a-b)^2}{(c+a)^2+b^2} + \frac{(a+b-c)^2}{(a+b)^2+c^2} \geq \frac{3}{5}$$

holds. Determine also when the equality holds.

Solution

Let

$$
\begin{aligned}
A &= (b+c-a)^2\left((c+a)^2+b^2\right)\left((a+b)^2+c^2\right) \\
&\quad +(c+a-b)^2\left((b+c)^2+a^2\right)\left((a+b)^2+c^2\right) \\
&\quad +(a+b-c)^2\left((b+c)^2+a^2\right)\left((c+a)^2+b^2\right), \\
B &= \left((b+c)^2+a^2\right)\left((c+a)^2+b^2\right)\left((a+b)^2+c^2\right).
\end{aligned}
$$

We have to prove that $5A - 3B \geq 0$.

We have

$$
\begin{aligned}
&5A - 3B \\
&= 4\{3(a^6+b^6+c^6) + (a^5b+ab^5+b^5c+bc^5+c^5a+ca^5) \\
&\quad - (a^4b^2+a^2b^4+b^4c^2+b^2c^4+c^4a^2+c^2a^4) \\
&\quad + 2(a^3b^3+b^3c^3+c^3a^3) + 3abc(a^3+b^3+c^3) \\
&\quad - 6abc(a^2b+ab^2+b^2c+bc^2+c^2a+ca^2) + 12a^2b^2c^2\} \\
&= 4\{bc(c-a)^2(a-b)^2 + ca(a-b)^2(b-c)^2 + ab(b-c)^2(c-a)^2\} \\
&\quad + 4abc\{(a+b)(a-b)^2 + (b+c)(b-c)^2 + (c+a)(c-a)^2\} \\
&\quad + 8\{a^3(b+c)(b-c)^2 + b^3(c+a)(c-a)^2 + c^3(a+b)(a-b)^2\} \\
&\quad + 6\{(a^3-b^3)^2 + (b^3-c^3)^2 + (c^3-a^3)^2\} \\
&\quad + 4\{ab(a^2+ab+b^2)(a-b)^2 + bc(b^2+bc+c^2)(b-c)^2 \\
&\quad + ca(c^2+ca+a^2)(c-a)^2\} \\
&\geq 0,
\end{aligned}
$$

with equality when $a = b = c$.

Let there be given a whole number $n > 1$, not divisible by 1997. Consider two sequences of numbers $\{a_i\}$ and $\{b_j\}$ defined by:

$$a_i = i + \frac{ni}{1997} \quad (i = 1, 2, 3, \ldots, 1996),$$

$$b_j = j + \frac{1997j}{n} \quad (j = 1, 2, 3, \ldots, n - 1).$$

By arranging the numbers of these two sequences in increasing order, we get the sequence $c_1 \leq c_2 \leq \cdots \leq c_{1995+n}$.

Prove that $c_{k+1} - c_k < 2$ for every $k = 1, 2, \ldots, 1994 + n$.

Solution

First note that $\{a_i\}$ and $\{b_j\}$ are two increasing arithmetical sequences, with difference $\alpha = 1 + \frac{n}{1997}$ and $\beta = 1 + \frac{1997}{n}$, respectively.

Let $i \in \{1, \ldots, 1996\}$ and $j \in \{1, \ldots, n - 1\}$, and suppose that $a_i = b_j$. Then $ni = 1997j$. Since 1997 is prime and $n \not\equiv 0 \pmod{1997}$, we deduce that $\gcd(n, 1997) = 1$. From Gauss's Theorem, we then have $i \equiv 0 \pmod{1997}$, which is impossible. It follows that $a_i \neq b_j$.

First Case. $n < 1997$.

We easily see that

$$\alpha < 2 < \beta, \tag{1}$$

$$\text{and} \quad a_1 < b_1, \tag{2}$$

which implies that $c_1 = a_1$. Moreover, $\frac{a_{1996}}{b_{n-1}} = \frac{1996n}{1997(n-1)} > 1$. Then,

$$b_{n-1} < a_{1996}, \tag{3}$$

which implies that $c_{1995+n} = a_{1996}$.

Lemma. For every $j \in \{1, \ldots, n - 2\}$, there exists $i \in \{1, \ldots, 1996\}$ such that $b_j < a_i < b_{j+1}$.

Proof. Suppose, for the purpose of contradiction, that there exists $j \in \{1, \ldots, n - 2\}$ such that the interval $[b_j, b_{j+1}]$ does not contain any of the a_i's. Let p be the greatest index such that $a_p < b_j$ (such a p does

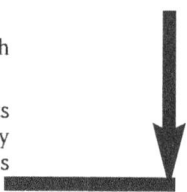

exist, since $a_1 < b_1 \leq b_j$). Then $p < 1996$ (since $b_j < b_{n-1} < a_{1996}$), and $a_p < b_j < b_{j+1} < a_{p+1}$. It follows that $\alpha = a_{p+1} - a_p > b_{j+1} - b_j = \beta$, which contradicts (1). Thus, the lemma is proved.

It follows from the lemma that, for every $k \in \{1, \ldots, 1994 + n\}$, we are in one of the three following cases:

(a) $c_k = a_i$ and $c_{k+1} = a_{i+1}$ for some i. Then $c_{k+1} - c_k = \alpha < 2$.

(b) $c_k = a_i$ and $c_{k+1} = b_j$ for some $i < 1996$ (from (3)) and some $j \leq n-1$. Then $b_j < a_{i+1}$ and $c_{k+1} - c_k < a_{i+1} - a_i = \alpha < 2$.

(c) $c_k = b_j$ and $c_{k+1} = a_i$ for some $i > 1$ (from (2)) and some $j \leq n - 1$. Then $a_{i-1} < b_j$ and $c_{k+1} - c_k < a_i - a_{i-1} = \alpha < 2$.

In each case, we have $c_{k+1} - c_k < 2$, as desired.

Second Case. $n > 1997$.

This case is essentially the same as the first case: simply interchange n with 1997 and a with b (and also α with β).

Prove that, for each prime number $p \geq 7$, there exists a positive integer n and integers $x_1, x_2, \ldots, x_n, y_1, y_2, \ldots, y_n$ which are not divisible by p, such that

$$
\begin{aligned}
x_1^2 + y_1^2 &\equiv x_2^2 \pmod{p} , \\
x_2^2 + y_2^2 &\equiv x_3^2 \pmod{p} , \\
&\vdots \\
x_{n-1}^2 + y_{n-1}^2 &\equiv x_n^2 \pmod{p} , \\
x_n^2 + y_n^2 &\equiv x_1^2 \pmod{p} .
\end{aligned}
$$

Solution

We consider two cases:

$$p \equiv 1 \pmod{4} \qquad \text{or} \qquad p \equiv 3 \pmod{4} .$$

(a) $p \equiv 3 \pmod{4}$. Then $p = 4k + 3$ for some integer k (where $k > 0$ since $p \geq 7$), and we observe that

$$1^2 + k^2 \equiv (k+2)^2 \pmod{p} . \tag{1}$$

Setting $x_1 = 1$, $y_1 = k$, and $x_2 = k + 2$, we have $x_1^2 + y_1^2 \equiv x_2^2 \pmod{p}$. Suppose now that we are given $x_i^2 + y_i^2 \equiv x_{i+1}^2 \pmod{p}$ for some $i \geq 1$. We will construct integers y_{i+1} and x_{i+2} such that $x_{i+1}^2 + y_{i+1}^2 \equiv x_{i+2}^2 \pmod{p}$. We first multiply (1) by x_{i+1}^2 to yield

$$x_{i+1}^2 + k^2 x_{i+1}^2 \equiv (k+2) x_{i+1}^2 \pmod{p} .$$

Then we choose $y_{i+1} \equiv k x_{i+1} \pmod{p}$ and $x_{i+2} \equiv (k+2) x_{i+1} \pmod{p}$.

Since, for any prime p, there are a finite number of quadratic residues, eventually we will have $x_j \equiv x_i \pmod{p}$ for some $j > i$. We can then re-label x_i as x_1 and y_i as y_1, and begin the process there.

For example, if $p = 13$, then $p = 4k + 1$ for $k = 3$. We start with

$$1^2 + 9^2 \equiv 11^2 \pmod{13} ,$$

and proceed to get the following circuit:

$$
\begin{aligned}
11^2 + 8^2 &\equiv 4^2 \pmod{13} , \\
4^2 + 10^2 &\equiv 5^2 \pmod{13} , \\
5^2 + 6^2 &\equiv 3^2 \pmod{13} , \\
3^2 + 1^2 &\equiv 7^2 \pmod{13} , \\
7^2 + 11^2 &\equiv 12^2 \equiv 1^2 \pmod{13} .
\end{aligned}
$$

(b) $p \equiv 1 \pmod{4}$. Then $p = 4k + 1$ for some integer k (where $k > 1$ since $p \geq 7$), and we observe that

$$1^2 + (3k)^2 \equiv (3k+2)^2 \pmod{p} . \tag{2}$$

Our process is similar to part (a), only this time we multiply (2) by x_{i+1}^2 and choose $y_{i+1} \equiv 3k x_{i+1} \pmod{p}$ and $x_{i+2} \equiv (3k+2) x_{i+1} \pmod{p}$.

Given an integer $n \geq 2$, find the minimal value of

$$\frac{x_1^5}{x_2 + x_3 + \cdots + x_n} + \frac{x_2^5}{x_1 + x_3 + \cdots + x_n} + \cdots + \frac{x_n^5}{x_1 + x_2 + \cdots + x_{n-1}}$$

subject to $x_1^2 + x_2^2 + \cdots + x_n^2 = 1$, where x_1, x_2, ..., x_n are positive real numbers.

Solution

More generally, we will find the minimal value of $\sum_{i=1}^{n} \left(\frac{x_i^p}{S - x_i^r} \right)$ subject to $x_1^s + x_2^s + \cdots + x_n^s = 1$, where $S = x_1^r + x_2^r + \cdots + x_n^r$ and p, r, s are positive real numbers such that $r \leq s \leq \frac{p}{2}$. The given problem is the special case $p = 5$, $r = 1$, $s = 2$.

Applying the Cauchy-Schwarz Inequality, we have

$$\left(\sum_{i=1}^{n} (S - x_i^r) \right) \left(\sum_{i=1}^{n} \frac{x_i^p}{S - x_i^r} \right) \geq \left(\sum_{i=1}^{n} x_i^{p/2} \right)^2,$$

with equality if and only if there exists a positive real number λ such that $S - x_i^r = \lambda x_i^{p/2}$.

By the Power-Mean Inequality (since $p/2 \geq s$),

$$\left(\sum_{i=1}^{n} x_i^{p/2} \right)^2 = n^2 \left(\sum_{i=1}^{n} \frac{x_i^{p/2}}{n} \right)^2 \geq n^2 \left(\sum_{i=1}^{n} \frac{x_i^s}{n} \right)^{p/s} = \frac{n^2}{n^{p/s}},$$

with equality if and only if $p = 2s$ or $x_1 = x_2 = \cdots = x_n = \frac{1}{n^{1/s}}$. Also,

$$\sum_{i=1}^{n} (S - x_i^r) = (n-1) \sum_{i=1}^{n} x_i^r = n(n-1) \sum_{i=1}^{n} \frac{x_i^r}{n}$$

$$\leq n(n-1) \left(\sum_{i=1}^{n} \frac{x_i^s}{n} \right)^{r/s} = \frac{n(n-1)}{n^{r/s}},$$

by the Power-Mean Inequality (since $s \geq r$). Here, equality occurs if and only if $r = s$ or $x_1 = x_2 = \cdots = x_n = \frac{1}{n^{1/s}}$. Since $\sum_{i=1}^{n} (S - x_i^r) > 0$, we have

$$\sum_{i=1}^{n} \frac{x_i^p}{S - x_i^r} \geq \frac{\left(\sum_{i=1}^{n} x_i^{p/2} \right)^2}{\sum_{i=1}^{n} (S - x_i^r)} \geq \frac{\dfrac{n^2}{n^{p/s}}}{\dfrac{n(n-1)}{n^{r/s}}} = \frac{n^{(r+s-p)/s}}{n-1}.$$

Equality occurs when $x_1 = x_2 = \cdots = x_n = \dfrac{1}{n^{1/s}}$. Therefore, the expression on the right side above is the minimal value of the sum on the left side, subject to $x_1^s + x_2^s + \cdots + x_n^s = 1$.

Setting $p = 5$, $r = 1$, and $s = 2$, we find that the minimal value in the given problem is $\dfrac{1}{n(n-1)}$.

Let x be a number such that

$$x + \frac{1}{x} = -1.$$

Compute

$$x^{1994} + \frac{-1}{x^{1994}}.$$

Solution

Let x be a number such that $x + \frac{1}{x} = -1$. Then $x = e^{2i\pi/3} = j$ or $x = \bar{j}$. Since $1/j = \bar{j}$ and $j^2 = \bar{j}$ and $j^3 = 1$, we deduce that

- If $x = j$, then $x^{1994} - \frac{1}{x^{1994}} = j^2 - \frac{1}{j^2} = \bar{j} - j = -i\sqrt{3}$.

- If $x = \bar{j}$, then $x^{1994} - \frac{1}{x^{1994}} = \overline{-i\sqrt{3}} = i\sqrt{3}$.

   Then

$$x^{1994} - \frac{1}{x^{1994}} = \pm i\sqrt{3}.$$

Let a be a natural number. Show that the equation

$$x^2 - y^2 = a^3$$

always has integer solutions for x and y.

Solution

Let $x = \dfrac{a(a+1)}{2}$, $y = \dfrac{a(1-a)}{2}$. Clearly, $x, y \in \mathbb{Z}$. Moreover,

$$
\begin{aligned}
x^2 - y^2 &= (x-y)(x+y) = \frac{a^2 + a - a + a^2}{2} \cdot \frac{a^2 + a + a - a^2}{2} \\
&= a^2 \cdot a = a^3.
\end{aligned}
$$

Solve the following system of equations in the set of complex numbers:

$$|z_1| = |z_2| = |z_3| = 1.$$
$$z_1 + z_2 + z_3 = 1.$$
$$z_1 z_2 z_3 = 1.$$

Solution

Let $\bar{z}$ denote the complex conjugate of z. We have $\bar{z}_i = 1/z_i$ for $i = 1, 2, 3$. It follows that

$$z_1 z_2 + z_2 z_3 + z_3 z_1 = \frac{1}{z_3} + \frac{1}{z_1} + \frac{1}{z_2} = \bar{z}_3 + \bar{z}_1 + \bar{z}_2 = \overline{z_1 + z_2 + z_3} = 1.$$

Consider the cubic polynomial

$$(x - z_1)(x - z_2)(x - z_3) = x^3 - x^2 + x - 1 = (x - 1)(x^2 + 1).$$

Since $1, \pm i$ are the roots we have that z_1, z_2, z_3 are equal to $1, i, -i$ in some order.

Given nonnegative real numbers $x_1, x_2, \ldots, x_k$ and positive integers k, m, n such that $km \leq n$, prove that

$$n\left(\prod_{i=1}^{k} x_i^m - 1\right) \leq m \sum_{i=1}^{k} (x_i^n - 1).$$

Solution

Let $P = x_1 x_2 \cdots x_k$. Since $\sum_{i=1}^{k} x_i^n \geq k P^{n/k}$, it suffices to show that

$$n P^m - n \leq mk P^{n/k} - mk$$

or that

$$\frac{P^m - 1}{m} \leq \frac{P^r - 1}{r} \tag{1}$$

where $r = n/k \geq m$. (1) follows from the known result that $(P^x - 1)/x$ is increasing in x for $x \geq 0$. A proof follows immediately from the integral representation

$$\frac{P^x - 1}{x \ln P} = \int_0^1 e^{xt \ln P} dt.$$

Finally it is to be noted that m and n need not be positive integers, just being positive reals with $n \geq km$ suffices.

Let $0 \leq x_i \leq 1$ and $x_i + y_i = 1$, for $i = 1, 2, \ldots, n$. Prove that

$$(1 - x_1 x_2 \cdots x_n)^m + (1 - y_1^m)(1 - y_2^m) \cdots (1 - y_n^m) \geq 1$$

for all positive integers m and n.

Solution

$$\prod_{i=1}^{n} \left(1 - \prod_{j=1}^{m} p_{ij}\right) + \prod_{j=1}^{m} \left(1 - \prod_{i=1}^{n} q_{ij}\right) > 1 \qquad (1)$$

where $p_{ij} + q_{ij} = 1$, $0 < p_{ij} < 1$, and m, n are positive integers greater than 1. (The cases $m, n = 1$ are trivial.) Just let $p_{ij} = x_j$ for all i and interchange m and n to get the given inequality. Other particularly nice special cases of the above inequality are

$$\left(1 - \frac{1}{2^n}\right)^m + \left(1 - \frac{1}{2^m}\right)^n > 1$$

and

$$\frac{1}{2^m} + \frac{1}{2^{1/m}} < 1.$$

If $\mathbf{a}$ and $\mathbf{b}$ are given nonparallel vectors, solve for x in the equation

$$\frac{a^2 + x\mathbf{a} \cdot \mathbf{b}}{|\mathbf{a}||\mathbf{a} + x\mathbf{b}|} = \frac{b^2 + \mathbf{a} \cdot \mathbf{b}}{|\mathbf{b}||\mathbf{a} + \mathbf{b}|}.$$

Solution

One can solve by squaring out and solving the rather messy quadratic in x. However, it is easier to proceed geometrically. Referring to the figure, the given equation requires that $\angle SPQ = \angle PRQ$ ($= \phi$, say). Thus $\triangle PQS \sim \triangle RQP$.

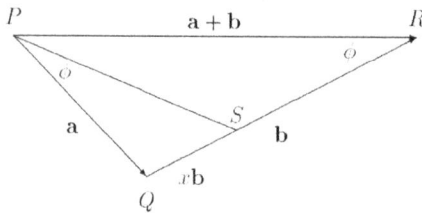

Thus, putting $a = |\mathbf{a}|$ and $b = |\mathbf{b}|$, $xb/a = a/b$ or $x = a^2/b^2$.

Also, $x = -x'$ can be negative as in the following figure.

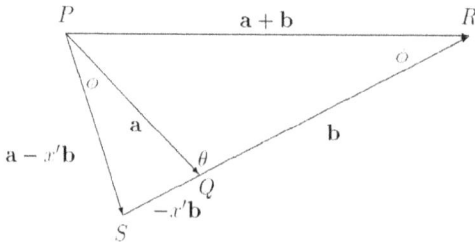

By the law of sines, $a/\sin(\theta - \phi) = x'b/\sin\phi$, and $a/\sin\phi = b/\sin(\theta + \phi)$. Hence,

$$\sin(\theta + \phi) - \sin(\theta - \phi) = \frac{b\sin\phi}{a} - \frac{a\sin\phi}{x'b}$$

or $2\cos\theta = b/a - a/x'b$. Finally,

$$x = -x' = \frac{-a^2}{b(b - 2a\cos\theta)} = \frac{-a^2}{b^2 + 2\mathbf{a} \cdot \mathbf{b}}.$$

Prove that

$$\frac{x_1^2}{x_1^2 + x_2 x_3} + \frac{x_2^2}{x_2^2 + x_3 x_4} + \cdots + \frac{x_{n-1}^2}{x_{n-1}^2 + x_n x_1} + \frac{x_n^2}{x_n^2 + x_1 x_2} \le n - 1.$$

where all $x_i > 0$.

Solution

Let S_n denote the cyclic sum. Clearly

$$S_2 = \frac{x_1^2}{x_1^2 + x_2 x_1} + \frac{x_2^2}{x_2^2 + x_1 x_2} = 1.$$

We prove by induction that $S_n < n - 1$ for all $n \ge 3$. Consider the case $n = 3$. Since the sum is cyclic and since the expression is unchanged when x_2 and x_3 are interchanged, we may assume that $x_1 \le x_2 \le x_3$. Then

$$S_3 = \frac{x_1^2}{x_1^2 + x_2 x_3} + \frac{x_2^2}{x_2^2 + x_3 x_1} + \frac{x_3^2}{x_3^2 + x_1 x_2} < \frac{x_1^2}{x_1^2 + x_2 x_1} + \frac{x_2^2}{x_2^2 + x_2 x_1} + 1 = 2.$$

Now suppose $S_n < n - 1$ for some $n \ge 3$ and consider S_{n+1} for $n + 1$ positive numbers $x_1, x_2, \ldots, x_{n+1}$. Without loss of generality, we may assume that $x_{n+1} = \max\{x_i : i = 1, 2, \ldots, n + 1\}$. Note that

$$S_{n+1} = S_n + \frac{x_{n-1}^2}{x_{n-1}^2 + x_n x_{n+1}} + \frac{x_n^2}{x_n^2 + x_{n+1} x_1} + \frac{x_{n+1}^2}{x_{n+1}^2 + x_1 x_2} - \frac{x_{n-1}^2}{x_{n-1}^2 + x_n x_1} - \frac{x_n^2}{x_n^2 + x_1 x_2}.$$

Since

$$\frac{x_{n-1}^2}{x_{n-1}^2 + x_n x_{n+1}} \le \frac{x_{n-1}^2}{x_{n-1}^2 + x_n x_1}, \quad \frac{x_n^2}{x_n^2 + x_{n+1} x_1} \le \frac{x_n^2}{x_n^2 + x_1 x_2}, \text{ and } \frac{x_{n+1}^2}{x_{n+1}^2 + x_1 x_2} < 1.$$

we conclude that $S_{n+1} < S_n + 1 < n$, completing the induction.

   Remark. If we set $x_i = t^i$ for $i = 1, 2, \ldots, n - 1$, and $x_n = t^{2n}$, then

$$S_n = \frac{t^2}{t^2 + t^5} + \frac{t^4}{t^4 + t^7} + \cdots + \frac{t^{2n-4}}{t^{2n-4} + t^{3n-1}} + \frac{t^{2n-2}}{t^{2n-2} + t^{2n+1}} + \frac{t^{4n}}{t^{4n} + t^3}$$

$$\longrightarrow n - 1 \text{ as } t \longrightarrow 0^+.$$

That is, though the bound $n - 1$ can not be attained for $n > 2$, it is nonetheless *sharp*.

Let $a_1, \ldots, a_{1988}$ be positive real numbers whose arithmetic mean equals 1988. Show

$$\sqrt[1988]{\prod_{i=1}^{1988} \prod_{j=1}^{1988} \left(1 + \frac{a_i}{a_j}\right)} \geq 2^{1988}$$

and determine when equality holds.

Solution

We shall prove the following general result which shows that the assumption about the arithmetic mean of the a_k's is redundant.

Theorem: If $a_i > 0$ for all $i = 1, 2, \ldots, n$ then

$$\left(\prod_{i=1}^{n} \prod_{j=1}^{n} \left(1 + \frac{a_i}{a_j}\right)\right)^{1/n} \geq 2^n$$

with equality if and only if the a_k's are equal.

Proof. It is well known (cf [1], [2]) and easy to show that if $a_1 a_2 \cdots a_n = b^n$, then

$$\prod_{i=1}^{n} (1 + a_i) \geq (1 + b)^n$$

with equality if and only if the a_k's are equal. Using this, we obtain for each fixed i,

$$\prod_{j=1}^{n} \left(1 + \frac{a_i}{a_j}\right) \geq \left(1 + \frac{a_i}{b}\right)^n$$

and hence

$$\left(\prod_{i=1}^{n} \prod_{j=1}^{n} \left(1 + \frac{a_i}{a_j}\right)\right)^{1/n} \geq \prod_{i=1}^{n} \left(1 + \frac{a_i}{b}\right) \geq \left(1 + \frac{b}{b}\right)^n = 2^n$$

Find all solutions (x, y, z) of the Diophantine equation

$$x^3 + y^3 + z^3 + 6xyz = 0.$$

Solution

The only nonzero solutions are $(x, y, z) = (1, -1, 0)$ in some order.

Mordell [1] has shown that the equation $x^3 + y^3 + z^3 + dxyz = 0$, for $d \neq 1, -3, -5$, has either three relatively prime solutions or an infinite number of solutions. Three of the solutions are $(1, -1, 0), (0, 1, -1)$ and $(-1, 0, 1)$. (Note that, e.g., $(1, -1, 0)$ and $(-1, 1, 0)$ are considered the same.) According to Dickson [2], Sylvester stated that $F \equiv x^3 + y^3 + z^3 + 6xyz$ is not solvable in integers (presumably non-trivially). The same holds for $2F = 27nxyz$ when $27n^2 - 8n + 4$ is a prime, and for $4F = 27nxyz$ when $27n^2 - 36n + 16$ is a prime.

In $\mathbf{R}^n$ let $\mathbf{X} = (x_1, x_2, \ldots, x_n)$, $\mathbf{Y} = (y_1, y_2, \ldots, y_n)$, and, for $p \in (0,1)$, define

$$F_p(\mathbf{X}, \mathbf{Y}) \equiv \left(\left|\frac{x_1}{p}\right|^p \left|\frac{y_1}{1-p}\right|^{1-p} \cdot \left|\frac{x_2}{p}\right|^p \left|\frac{y_2}{1-p}\right|^{1-p} \cdot \ldots \cdot \left|\frac{x_n}{p}\right|^p \left|\frac{y_n}{1-p}\right|^{1-p} \right).$$

Prove that

$$\|\mathbf{X}\|_m + \|\mathbf{Y}\|_m \ge \|F_p(\mathbf{X}, \mathbf{Y})\|_m.$$

where

$$\|\mathbf{X}\|_m = (|x_1|^m + |x_2|^m + \cdots + |x_n|^m)^{1/m}.$$

Solution

By Hölder's inequality

$$\|F_p(\mathbf{X}, \mathbf{Y})\| \le \frac{\|\mathbf{X}\|_m^p \|\mathbf{Y}\|_m^{1-p}}{p^p(1-p)^{1-p}}.$$

By the weighted A.M.-G.M. inequality,

$$\|\mathbf{X}\|_m + \|\mathbf{Y}\|_m = p\left\|\frac{\mathbf{X}}{p}\right\|_m + (1-p)\left\|\frac{\mathbf{Y}}{1-p}\right\|_m \ge \frac{\|\mathbf{X}\|_m^p \|\mathbf{Y}\|_m^{1-p}}{p^p(1-p)^{1-p}}.$$

The result follows.

If $m \ge 1$, and $x_i, y_i > 0$ for all i, then another proof, via Minkowski's inequality and the weighted A.M.-G.M. inequality, is

$$\|\mathbf{X}\|_m + \|\mathbf{Y}\|_m \ge \|\mathbf{X} + \mathbf{Y}\|_m \ge \|F_p(\mathbf{X}, \mathbf{Y})\|_m.$$

Determine all values of the real parameter p for which the system of equations

$$x + y + z = 2$$

$$yz + zx + xy = 1$$

$$xyz = p$$

has a real solution.

Solution

Since
$$x = xyz + x^2 z + x^2 y = p + x^2(2 - x) = p + 2x^2 - x^3,$$
the solution (x, y, z) is given by the three roots, in any order, of the cubic equation

$$t^3 - 2t^2 + t - p = 0.$$

As is known [1], the condition that the general cubic equation

$$at^3 + 3bt^2 + 3ct + d = 0$$

have real roots is that $\Delta \leq 0$ where

$$\Delta = a^2 d^2 - 6abcd + 4ac^3 + 4db^3 - 3b^2 c^2.$$

For the case here.
$$\Delta = p^2 - 4p/27.$$

Consequently, p must lie in the closed interval $[0, 4/27]$.

The real numbers $x_1, x_2, x_3, \ldots$ are defined by

$$x_1 = a \neq -1 \quad \text{and} \quad x_{n+1} = x_n^2 + x_n \text{ for all } n \geq 1.$$

S_n is the sum and P_n is the product of the first n terms of the sequence $y_1, y_2, y_3, \ldots,$ where

$$y_n = \frac{1}{1 + x_n}.$$

Prove that $aS_n + P_n = 1$ for all n.

Solution

It follows easily that

$$\frac{1}{y_{n+1}} = \frac{1}{y_n^2} - \frac{1}{y_n} + 1. \tag{1}$$

where $y_1 = 1/(1 + a)$. Now let

$$\varphi_n = aS_{n+1} + P_{n+1} - aS_n - P_n = P_n(y_{n+1} - 1) + ay_{n+1}. \tag{2}$$

Replacing n by $n - 1$ and dividing gives

$$\frac{y_n(y_{n+1} - 1)}{y_n - 1} = \frac{\varphi_n - ay_{n+1}}{\varphi_{n-1} - ay_n}.$$

It follows from (1) that

$$\frac{y_n(y_{n+1} - 1)}{y_n - 1} = \frac{y_{n+1}}{y_n}.$$

Hence

$$\frac{\varphi_n - ay_{n+1}}{\varphi_{n-1} - ay_n} = \frac{y_{n+1}}{y_n}$$

or

$$\varphi_n y_n = \varphi_{n-1} y_{n+1}. \tag{3}$$

An easy calculation shows that $y_2 = 1/(1 + a + a^2)$, so that

$$\varphi_1 = P_1(y_2 - 1) + ay_2 = \frac{1}{1 + a}\left(\frac{1}{1 + a + a^2} - 1\right) + \frac{a}{1 + a + a^2}$$

$$= \frac{1 + a(1 + a)}{(1 + a)(1 + a + a^2)} - \frac{1}{1 + a} = 0.$$

Then since $y_i \neq 0$ for all i, from (3) $\varphi_i = 0$ for all i. Since $aS_1 + P_1 = 1$, by (2) $aS_n + P_n = 1$ for all n.

Let $x_1, x_2, \ldots, x_n$ be positive real numbers, and let $S = x_1 + x_2 + \cdots + x_n$. Prove that

$$(1 + x_1)(1 + x_2) \cdots (1 + x_n) \leq 1 + S + \frac{S}{2!} + \frac{S}{3!} + \cdots + \frac{S}{n!}.$$

Solution

The correct inequality should be

$$(1 + x_1)(1 + x_2) \cdots (1 + x_n) \leq 1 + S + \frac{S^2}{2!} + \cdots + \frac{S^n}{n!}.$$

$$(1 + x_1) \cdots (1 + x_n) \leq \left(\frac{n + x_1 + \ldots + x_n}{n} \right)^n = \left(1 + \frac{S}{n} \right)^n$$

$$= 1 + n \left(\frac{S}{n} \right) + \frac{n(n-1)}{2} \left(\frac{S}{n} \right)^2 + \cdots + \left(\frac{S}{n} \right)^n$$

using the binomial theorem. Since $(n - m)! n^m \geq n!$, the coefficient of S^m is

$$\binom{n}{m} \frac{1}{n^m} = \frac{n!}{m!(n - m)! n^m} \leq \frac{n!}{m! n!} = \frac{1}{m!},$$

from which the result is immediate.

Prove that, for all natural numbers $n \geq 2$,

$$\prod_{i=1}^{n} \tan\left\{\frac{\pi}{3}\left(1 + \frac{3^i}{3^n - 1}\right)\right\} = \prod_{i=1}^{n} \cot\left\{\frac{\pi}{3}\left(1 - \frac{3^i}{3^n - 1}\right)\right\} .$$

Solution

Let

$$A_i = \tan\left\{\frac{\pi}{3}\left(1 + \frac{3^i}{3^n - 1}\right)\right\} = \frac{\tan\frac{\pi}{3} + \tan(\frac{\pi 3^{i-1}}{3^n-1})}{1 - \tan\frac{\pi}{3}\tan(\frac{\pi 3^{i-1}}{3^n-1})}$$

and

$$B_i = \tan\left\{\frac{\pi}{3}\left(1 - \frac{3^i}{3^n - 1}\right)\right\} = \frac{\tan\frac{\pi}{3} - \tan(\frac{\pi 3^{i-1}}{3^n-1})}{1 + \tan\frac{\pi}{3}\tan(\frac{\pi 3^{i-1}}{3^n-1})}.$$

Now since

$$\tan 3\theta = \tan\theta\left[\frac{3 - \tan^2\theta}{1 - 3\tan^2\theta}\right]$$

we get

$$A_i B_i = \frac{3 - \tan^2(\frac{\pi 3^{i-1}}{3^n-1})}{1 - 3\tan^2(\frac{\pi 3^{i-1}}{3^n-1})} = \frac{\tan(\frac{\pi 3^i}{3^n-1})}{\tan(\frac{\pi 3^{i-1}}{3^n-1})}.$$

Hence

$$\prod_{i=1}^{n} A_i B_i = \frac{\tan(\frac{\pi 3^n}{3^n-1})}{\tan(\frac{\pi}{3^n-1})} = \frac{\tan(\pi + \frac{\pi}{3^n-1})}{\tan\left(\frac{\pi}{3^n-1}\right)} = 1.$$

From this the result is immediate.

Determine all real solutions x, y of the system

$$x^4 + y^2 - xy^3 - 9x/8 = 0,$$
$$y^4 + x^2 - yx^3 - 9y/8 = 0.$$

Solution

We find instead all the solutions of the given equations, which we rewrite as

$$8x^4 + 8y^2 - 8xy^3 - 9x = 0 \ . \tag{1}$$
$$8y^4 + 8x^2 - 8yx^3 - 9y = 0 \ . \tag{2}$$

We show there are exactly *ten* real or complex solutions, namely

$$(0,0), (\frac{9}{8}, \frac{9}{8}), (1, \frac{1}{2}), (\frac{1}{2}, 1), (\omega, \frac{\omega^2}{2}), (\omega^2, \frac{\omega}{2})(\frac{\omega}{2}, \omega^2), (\frac{\omega^2}{2}, \omega), (\frac{9\omega}{8}, \frac{9\omega^2}{8}), (\frac{9\omega^2}{8}, \frac{9\omega}{8}).$$

where $\omega = (-1 + \sqrt{3}i)/2$ denotes a complex cube root of unity.

From (y times (1)) minus (x times (2)) we get

$$0 = 2(x^4 y - xy^4) - (x^3 - y^3) = (x^3 - y^3)(2xy - 1)$$
$$= (x - y)(2xy - 1)(x^2 + xy + y^2) \ .$$

If $x = y$, then substitution in (1) yields $8x^2 - 9x = 0$ which immediately gives two candidates $(0,0)$ and $(9/8, 9/8)$. If $2xy = 1$, then substituting $y = (2x)^{-1}$ in (1) and simplifying we obtain

$$8x^4 + \frac{1}{x^2} - 9x = 0$$

or $8x^6 - 9x^3 + 1 = 0$. Setting $t = x^3$ we obtain $8t^2 - 9t + 1 = 0$ which yields $t = 1, 1/8$. These give six more candidates

$$(1, \frac{1}{2}), (\omega, \frac{\omega^2}{2}), (\omega^2, \frac{\omega}{2}), (\frac{1}{2}, 1), (\frac{\omega}{2}, \omega^2), (\frac{\omega^2}{2}, \omega).$$

Finally, suppose

$$x^2 + xy + y^2 = 0 \ . \tag{3}$$

From (1) minus (2) we obtain

$$8(x^4 - y^4) - 8(x^2 - y^2) + 8xy(x^2 - y^2) - 9(x - y) = 0 \ .$$

and disregarding the possibility that $x = y$, a case already considered above, we have

$$(x + y)[8(x^2 + y^2) - 8 + 8xy] = 9 \ . \tag{4}$$

Substitution of (3) into (4) now yields

$$x + y = -\frac{9}{8} \ . \tag{5}$$

From (5) and (3) we get

$$xy = (x + y)^2 - (x^2 + xy + y^2) = \frac{81}{64} \ .$$ (6)

From (5) and (6) we see that x and y are the two roots of the equation

$$u^2 + \frac{9}{8}u + \frac{81}{64} = 0 \ .$$

or $64u^2 + 72u + 81 = 0$. Solving. we get $u = (-9 \pm 9\sqrt{3}i)/16$. This gives two more (complex) candidates

$$(\frac{9\omega}{8}, \frac{9\omega^2}{8}). \ (\frac{9\omega^2}{8}, \frac{9\omega}{8}).$$

Straightforward substitution into (1) and (2) shows that all of these ten candidates are solutions.

Prove that for any natural number k, there is an integer n such that

$$\sqrt{n + 1981^k} + \sqrt{n} = (\sqrt{1982} + 1)^k .$$

Solution

Let

$$A = \sum_{\substack{j=0 \\ j \text{ even}}}^{k} \binom{k}{j}(\sqrt{1982})^j , \quad B = \sum_{\substack{j=0 \\ j \text{ odd}}}^{k} \binom{k}{j}(\sqrt{1982})^j .$$

[Thus A is the sum of the even-numbered terms in the expansion of $(\sqrt{1982} + 1)^k$ and B is the sum of the odd-numbered terms in that expansion.] Note also that

$$(\sqrt{1982} - 1)^k = \sum_{j=0}^{k} \binom{k}{j}(\sqrt{1982})^{k-j}(-1)^j = \begin{cases} B - A & \text{if } k \text{ is odd} \\ A - B & \text{if } k \text{ is even} . \end{cases}$$

Case 1: k is odd. Let $n = A^2$. Then

$$\begin{aligned}
\sqrt{n + 1981^k} + \sqrt{n} &= \sqrt{A^2 + (\sqrt{1982} - 1)^k(\sqrt{1982} + 1)^k} + A \\
&= \sqrt{A^2 + (B - A)(A + B)} + A \\
&= \sqrt{A^2 + B^2 - A^2} + A \\
&= B + A = (\sqrt{1982} + 1)^k.
\end{aligned}$$

Case 2: k is even. Let $n = B^2$. Then

$$\begin{aligned}
\sqrt{n + 1981^k} + \sqrt{n} &= \sqrt{B^2 + (A - B)(A + B)} + \sqrt{B^2} \\
&= \sqrt{A^2} + \sqrt{B^2} = (\sqrt{1982} + 1)^k.
\end{aligned}$$

It is evident that A^2 is an integer, and B^2 is an integer since B has the form $\sqrt{1982}\, L$, for some integer L.

Let a, b, c, d, p and q be natural numbers different from zero such that

$$ad - bc = 1 \quad \text{and} \quad \frac{a}{b} > \frac{p}{q} > \frac{c}{d}.$$

Show that

(i) $q \geq b + d$;

(ii) if $q = b + d$ then $p = a + c$.

Solution

Since $a/b > p/q$, $aq - pd > 0$, so $aq - pd \geq 1$. Likewise $pd - cq \geq 1$. Now

$$q = q(ad - bc) = b(pd - cq) + d(aq - pb) \geq b + d.$$

If $q = b + d$, then $b(pd - cq - 1) + d(ap - bp - 1) = 0$. Since $b > 0$, $d > 0$, $pd - cq - 1 \geq 0$ and $aq - bp - 1 \geq 0$ we must have $pd - cq = aq - pb = 1$. This yields $q(a + c) = (d + b)p$. But $q = b + d$, so $a + c = p$ as required.

We consider expressions of the form

$$x + yt + zt^2 ,$$

where $x, y, z \in \mathbf{Q}$, and $t^2 = 2$. Show that, if $x + yt + zt^2 \neq 0$, then there exist $u, v, w \in \mathbf{Q}$ such that

$$(x + yt + zt^2)(u + vt + wt^2) = 1 .$$

Solution

Observe that $(x + yt + zt^2)(x + zt^2 - yt) = (x + 2z)^2 - 2y^2$. Now $(x + 2z)^2 - 2y^2 \neq 0$, for otherwise $\sqrt{2} = |(x + 2z)/y|$ is rational. [Note if $y = 0$ in this case, then $x + 2z = 0$ and $x + yt + zt^2 = 0$, contrary to assumption.] Let $\alpha = (x + 2z)^2 - 2y^2$. Let $u = x/\alpha$, $v = -y/\alpha$ and $w = z/\alpha$. Then

$$(x + yt + zt^2)(u + vt + wt^2) = \frac{(x + 2y)^2 - 2y^2}{\alpha} = 1 .$$

Determine all the real roots of $4x^4 + 16x^3 - 6x^2 - 40x + 25 = 0$.

Solution

Dividing by x^2 we get $4x^2 + 16x - 6 - 40/x + 25/x^2 = 0$. The substitution $z = 2x - 5/x$ yields $z^2 + 8z + 14 = 0$. This gives solutions $z_1 = -4 + \sqrt{2}$ and $z_2 = -4 - \sqrt{2}$.

Case 1. $2x^2 - (-4 + \sqrt{2})x - 5 = 0$. Then

$$x = \frac{-4 + \sqrt{2} \pm \sqrt{58 - 8\sqrt{2}}}{4} .$$

Case 2. $2x^2 + (4 + \sqrt{2})x - 5 = 0$. Then

$$x = \frac{-4 - \sqrt{2} \pm \sqrt{58 + 8\sqrt{2}}}{4} .$$

This gives the four real roots.

Determine the value of p such that the equation $x^5 - px - 1 = 0$ has two roots r and s which are the roots of an equation $x^2 - ax + b = 0$ where a and b are integers.

Solution

If p satisfies the condition

$$x^5 - px - 1 = (x^2 - ax + b)(x^3 + ux^2 + vx - 1/b).$$

comparing coefficients of x^4 we get $u - a = 0$, so $u = a$. From coefficients of x^3 and x^2 we then get, respectively,

$$b - a^2 + v = 0 \quad \text{and} \quad \frac{-1}{b} - av + ab = 0. \tag{1}$$

and from coefficients of x,

$$\frac{a}{b} + bv = -p. \tag{2}$$

From (1),

$$\frac{-1}{b} - a(a^2 - b) + ab = 0.$$

Since a and b are integers it follows that b must be ± 1.

Now, $b = -1$ gives $a^3 + 2a - 1 = 0$. Since $(-1)^3 + 2(-1) - 1 = -4$ and $1^3 + 2 \cdot 1 - 1 = 2$, this gives no rational and hence no integer solutions for a. The case $b = 1$ gives $a^3 - 2a + 1 = 0$ which gives $a = 1$. From (1) $v = 0$ and so from (2) $p = -1$.

Let a, b, c, d, m, n be positive integers such that

$$a^2 + b^2 + c^2 + d^2 = 1989,$$
$$a + b + c + d = m^2,$$

and the largest of a, b, c, d is n^2. Determine, with proof, the values of m and n.

Solution

Without loss of generality we may suppose $0 < a \leq b \leq c \leq d$. Let $S = a^2 + b^2 + c^2 + d^2$. Since $x^2 + y^2 \geq 2xy$ it follows that

$$3S \geq 2(ab + ac + ad + bc + bd + cd)$$

whence

$$4S \geq (a + b + c + d)^2.$$

Thus $4 \cdot 1989 \geq m^4$ and $m \leq 9$.

But $(a+b+c+d)^2 > a^2+b^2+c^2+d^2$ so $m^4 > 1989$ and $m > 6$. Since $a^2+b^2+c^2+d^2$ is odd, so is $m^2 = a + b + c + d$, thus $m = 7$ or $m = 9$. Suppose for a contradiction that $m = 7$. Now $(49 - d)^2 = (a + b + c)^2 > a^2 + b^2 + c^2 = 1989 - d^2$. Thus $d^2 - 49d + 206 > 0$. It follows that $d > 44$ or $d \leq 4$. However, $d < 45$, since $45^2 = 2025$. Also, $d \leq 4$ implies $a^2 + b^2 + c^2 + d^2 \leq 64 < 1989$.

It follows that $m = 9$. Now $n^2 = d > 16$, since $d \leq 16$ implies $a+b+c+d \leq 64 < 81$. As before $d \leq 44$ so $n^2 = 25$ or 36. If $d = n^2 = 25$ then let $a = 25 - p$, $b = 25 - q$, $c = 25 - r$, with $p, q, r \geq 0$. Furthermore $a + b + c = 56$ implies $p + q + r = 19$. $(25-p)^2 + (25-q)^2 + (25-r)^2 = 1364$ implies $p^2 + q^2 + r^2 = 439$. Now $(p+q+r)^2 > p^2 + q^2 + r^2$ gives a contradiction.

Thus the only possibility is that $n = 6$, and there is a solution with $a = 12$, $b = 15$, $c = 18$, $d = 36$. Thus $m = 9$ and $n = 6$.

Show that the equation $x^2 + y^5 = z^3$ has infinitely many solutions in integers x, y, z for which $xyz \neq 0$.

Solution

Two solutions found by inspection are $(x, y, z) = (3, -1, 2)$ and $(10, 3, 7)$.
Suppose a solution $(x, y, z) = (u, v, w)$ is given. Then for any positive integer k, $(x, y, z) = (k^{15}u, k^6 v, k^{10}w)$ is also a solution.

Prove that the equation

$$6(6a^2 + 3b^2 + c^2) = 5n^2$$

has no solution in integers except $a = b = c = n = 0$.

Solution

We must have that $6|n$, and then $3|c$. Hence $2a^2 + b^2 + 3d^2 = 10m^2$ where $n = 6m$ and $c = 3d$. If the original equation has a non-trivial solution, then this equation has one with $\gcd(a, b, d, m) = 1$. Clearly $b \equiv d \bmod 2$. Now the quadratic residues modulo 16 are 0, 1, 4 and 9. The following table gives the possibilities for $2a^2 + b^2 + 3d^2 \bmod 16$, depending on the parities of a, b and d:

	b and d odd $b^2 \equiv 1$ or 9, $3d^2 \equiv 3$ or $11 \bmod 16$	b and d even $b^2 \equiv 0$ or 4, $3d^2 \equiv 0$ or $12 \bmod 16$
a odd $2a^2 \equiv 2 \bmod 16$	$2a^2 + b^2 + 3d^2 \equiv 6$ or $14 \bmod 16$	$2a^2 + b^2 + 3d^2 \equiv 2, 6$ or $14 \bmod 16$
a even $2a^2 \equiv 0$ or $8 \bmod 16$	$2a^2 + b^2 + 3d^2 \equiv 4$ or $12 \bmod 16$	$2a^2 + b^2 + 3d^2 \equiv 0, 4, 8$ or $12 \bmod 16$

If m is odd then $10m^2 \equiv 10 \bmod 16$, and we cannot have $2a^2 + b^2 + 3d^2 = 10m^2$. Hence m is even so that $10m^2 \equiv 0$ or $8 \bmod 16$. From the above table, we must have a, b, d all even. However this contradicts that $\gcd(a, b, d, m) = 1$. It follows that $6(6a^2 + 3b^2 + c^2) = 5n^2$ has no solution in integers except $a = b = c = n = 0$.

Determine the maximum value of

$$x^3 + y^3 + z^3 - x^2 y - y^2 z - z^2 x$$

for $0 \le x, y, z \le 1$.

Solution

Let $f(x, y, z)$ denote the given expression. We show that $f(x, y, z) \le 1$ with equality if and only if (x, y, z) equals one of the six triples: $(1, 0, 0)$, $(0, 1, 0)$, $(0, 0, 1)$, $(1, 1, 0)$, $(1, 0, 1)$ and $(0, 1, 1)$.

Suppose first that $0 \le x \le y \le z \le 1$. Then

$$1 - f(x, y, z) = (1 - z^3) + x^2(y - x) + y^2(z - y) + z^2 x \ge 0$$

with equality if and only if (i) $z = 1$, (ii) $x = 0$ or $y = x$, (iii) $y = 0$ or $z = y$, and (iv) $z = 0$ or $x = 0$. From (i) and (iv) we get $x = 0$, $z = 1$ which together with (ii) and (iii) immediately yields $(x, y, z) = (0, 0, 1)$ or $(0, 1, 1)$.

Next suppose that $0 \le y \le x \le z \le 1$. Then

$$1 - f(x, y, z) = (1 - z^3) + x(z^2 - x^2) + y(x^2 - y^2) + y^2 z \ge 0$$

with equality just when (i) $z = 1$, (ii) $x = 0$ or $x = z$, (iii) $y = 0$ or $x = y$, and (iv) $y = 0$ or $z = 0$. From (i) and (iv) we conclude that $z = 1$, $y = 0$. Now (ii) gives the two solutions $(x, y, z) = (0, 0, 1)$ and $(1, 0, 1)$.

Exploiting the cyclic symmetry of $f(x, y, z)$ we may reduce to one of these two cases, and this gives the six solutions listed.

Determine all quadruples (x, y, u, v) of real numbers satisfying the simultaneous equations

$$x^2 + y^2 + u^2 + v^2 = 4,$$

$$xu + yv = -xv - yu,$$

$$xyu + yuv + uvx + vxy = -2,$$

$$xyuv = -1.$$

Solution

The second equation can be rewritten

$$(x + y)(u + v) = 0$$

and the third is $xy(u + v) + (x + y)uv = -2$.

Case 1. $x + y = 0$. Then the last two equations give $x^2(u + v) = 2$ and $x^2 uv = 1$. From this $u + v = 2uv$ and $x^2 = y^2 = 2/(u + v)$. Using the first equation,

$$\frac{4}{u + v} + (u + v)^2 - (u + v) = 4.$$

Setting $z = u + v$ we obtain

$$(z - 1)(z - 2)(z + 2) = z^3 - z^2 - 4z + 4 = 0.$$

This now gives three subcases.

Subcase (i): $u + v = 1 = 2uv$, no real solution.

Subcase (ii): $u + v = 2 = 2uv$, giving $u = 1$, $v = 1$, $x = \pm 1$, $y = \pm 1$ and solutions $(1, -1, 1, 1)$, $(-1, 1, 1, 1)$.

Subcase (iii): $u + v = -2 = 2uv$, and no real solution for x.

Case 2. $u + v = 0$. By symmetry we obtain $(1, 1, -1, 1)$ and $(1, 1, 1, -1)$.

Find all positive integers x, y, z satisfying the equation $5(xy + yz + zx) = 4xyz$.

Solution

The given equation can be rewritten as

$$\frac{1}{x} + \frac{1}{y} + \frac{1}{z} = \frac{4}{5} \qquad (*)$$

Without loss of generality, we may assume that $1 \leq x \leq y \leq z$. Since x, y and z are positive $x = 1$ is clearly impossible. On the other hand, if $x \geq 4$, then

$$\frac{1}{x} + \frac{1}{y} + \frac{1}{z} \leq \frac{3}{4} < \frac{4}{5}.$$

Thus $(*)$ can have integer solutions only if $x = 2$ or $x = 3$.

When $x = 3$, $(*)$ becomes $\frac{1}{y} + \frac{1}{z} = \frac{7}{15}$. If $y \geq 5$, then $\frac{1}{y} + \frac{1}{z} \leq \frac{2}{5} < \frac{7}{15}$. Thus $y = 3$ or 4. In either case, we can easily find that the corresponding value for z is not an integer.

When $x = 2$, $(*)$ becomes $\frac{1}{y} + \frac{1}{z} = \frac{3}{10}$. If $y \leq 3$, then $\frac{1}{y} + \frac{1}{z} > \frac{1}{3} > \frac{3}{10}$. If $y \geq 7$ then $\frac{1}{y} + \frac{1}{z} \leq \frac{2}{7} < \frac{3}{10}$. Thus $y = 4, 5,$ or 6.

For $y = 4$, we solve and get $z = 20$, and a solution $(2, 4, 20)$. For $y = 5$, we have $z = 10$ and the solution $(2, 5, 10)$. For $y = 6$ we find $z = 15/2$, which is not an integer.

To summarize, the given equation has exactly 12 solutions obtained by permuting the entries of each of the two ordered triples $(2, 4, 20)$ and $(2, 5, 10)$.

Find the sum of the infinite series

$$1 + \frac{1}{2} + \frac{1}{3} + \frac{1}{4} + \frac{1}{6} + \frac{1}{8} + \frac{1}{9} + \frac{1}{12} + \cdots$$

where the terms are reciprocals of integers divisible only by the primes 2 or 3.

Solution

One must add "except for the first term 1" at the end of the statement.

If we replace reciprocals of the numbers of the form $2^k 3^\ell$ with reciprocals of numbers of the form $P_1^{k_1} P_2^{k_2} \ldots P_n^{k_n}$ where $P_1, P_2, \ldots, P_n$ are distinct primes, then the sum is given by

$$\left(\sum_{k=0}^{\infty} \frac{1}{P_1^k} \right) \left(\sum_{k=0}^{\infty} \frac{1}{P_2^k} \right) \cdots \left(\sum_{k=0}^{\infty} \frac{1}{P_n^k} \right) = \frac{P_1}{P_1 - 1} \cdot \frac{P_2}{P_2 - 1} \cdot \ldots \cdot \frac{P_n}{P_n - 1} \, .$$

For the given problem this gives $\frac{2}{1} \cdot \frac{3}{2} = 3$.

Let a and b be positive integers such that $ab + 1$ divides $a^2 + b^2$. Show that

$$\frac{a^2 + b^2}{ab + 1}$$

is the square of an integer.

Solution

If (a_i) is a doubly-infinite sequence (i.e., $-\infty < i < \infty$) of reals, satisfying

(1) $a_{i+2} = na_{i+1} - a_i$ for all i, and

(2) $\dfrac{a_i^2 + a_{i+1}^2}{a_i a_{i+1} + 1} = n$ for *one* value of i,

then (2) is satisfied for *all* values of i.

To see it, observe that

$$\frac{a_{i+2}^2 + a_{i+1}^2}{a_{i+2} a_{i+1} + 1} = \frac{(na_{i+1} - a_i)^2 + a_{i+1}^2}{(na_{i+1} - a_i)a_{i+1} + 1} = \frac{(a_{i+1}^2 + a_i^2) + n(na_{i+1}^2 - 2a_{i+1}a_i)}{(a_{i+1}a_i + 1) + (na_{i+1}^2 - 2a_{i+1}a_i)} = n$$

where the last equality uses the assumption (2). Observe that $a_{i+2} = na_{i+1} - a_i$ is equivalent to $a_i = na_{i+1} - a_{i+2}$, thus it follows that (2) holds for all i.

To solve the problem, suppose a and b are integers, for which $(a^2 + b^2)/(ab + 1) = n$ is an integer. If $a = b$, then it follows easily that $a = b = 1$, and the assertion is elementary. Otherwise, say $1 < a < b$, and let $a_1 = a$ and $a_2 = b$, and define (a_i) by $a_{i+2} = na_{i+1} - a_i$ and by $a_i = na_{i+1} - a_{i+2}$. It follows easily (since $n > 1$) that the doubly infinite sequence of integers a_i is monotonic increasing (in i), thus in going backwards with i, a_i tends to $-\infty$. Thus a_i becomes negative for suitable i. However, since n is positive, there is no j for which $a_j > 0$ and $a_{j-1} < 0$, because of (2). Thus $a_{j-1} = 0$ for some j, implying (use (2) with $i = j - 1$) that $n = a_j^2$.

As a by-product, we can get all the triples $(a, b, n) \neq (1, 1, 1)$ of integers satisfying $(a^2 + b^2)/(ab + 1) = n$. Such a and b will be any two consecutive terms of the sequence (a_i) satisfying $a_0 = 0$, $a_1 = m$ and defined by (1) where $n = m^2$. The characteristic equation of (1) is $x^2 - m^2 x + 1 = 0$, which has the solutions

$$x = \frac{m^2 \pm \sqrt{m^4 - 1}}{2};$$

thus, for suitable α and β,

$$a_i = \alpha \left(\frac{m^2 + \sqrt{m^4 - 4}}{2} \right)^i + \beta \left(\frac{m^2 - \sqrt{m^4 - 4}}{2} \right)^i$$

for all $i \geq 0$. From $a_0 = 0$ and $a_1 = m$ we get

$$a_i = \frac{m}{\sqrt{m^4 - 4}} \left(\frac{m^2 + \sqrt{m^4 - 4}}{2} \right)^i - \frac{m}{\sqrt{m^4 - 4}} \left(\frac{m^2 - \sqrt{m^4 - 4}}{2} \right)^i$$

and the general solution is (a_i, a_{i+1}, m^2) for $i = 0, 1, \ldots$.

Does the equation $x^2 + y^3 = z^4$ have solutions for prime numbers x, y and z?

Solution

We shall show that the equation has *no* solutions for a positive integer x and prime numbers y and z.

Suppose on the contrary that there is a solution of this form. The equation is equivalent to

$$y^3 = (z^2 - x)(z^2 + x).$$

Because y is a prime number and x is a positive integer, $z^2 - x = 1$ and $z^2 + x = y^3$ or $z^2 - x = y$ and $z^2 + x = y^2$.

If $z^2 - x = 1$ and $z^2 + x = y^3$, then

$$2z^2 = y^3 + 1 = (y+1)(y^2 - y + 1).$$

It follows that $y \neq 2$, so that y is an odd prime number. Thus $y > 2$, making $y + 1 > 3$ and $y^2 - y + 1 > y + 1$. Since z is a prime number we now obtain $z = y + 1$ and $y^2 - y + 1 = 2z$. But $z = y + 1$ where y and z are prime numbers with $y > 2$ is impossible.

If $z^2 - x = y$ and $z^2 + x = y^2$ then

$$2z^2 = y^2 + y = y(y + 1).$$

It follows that $y \neq 2$, so y is an odd prime. Thus $y > 2$. Since $y < y + 1$ and z is a prime we get $y = z$ and $y + 1 = 2z$. This gives $z = 1$, a contradiction.

Prove that the inequality

$$\sum_{n=1}^{r} \left(\sum_{m=1}^{r} \frac{a_m a_n}{m+n} \right) \geq 0$$

holds for any real numbers $a_1, a_2, \ldots, a_r$. Find conditions for equality.

Solution

Consider the polynomial

$$p(x) = \sum_{n=1}^{r} \left(\sum_{m=1}^{r} a_m a_n x^{m+n-1} \right).$$

Then

$$xp(x) = \sum_{n=1}^{r} \sum_{m=1}^{r} a_m a_n x^{m+n} = \left(\sum_{m=1}^{r} a_m x^m \right) \left(\sum_{n=1}^{r} a_n x^n \right)$$

$$= \left(\sum_{i=1}^{r} a_i x^i \right)^2 \geq 0,$$

for all $x \in \mathbb{R}$.

In particular $p(x) \geq 0$ for all $x \geq 0$. Hence

$$0 \leq \int_0^1 p(x)\, dx = \sum_{n=1}^{r} \left(\sum_{m=1}^{r} \frac{a_m a_n}{m+n} x^{m+n} \Big]_0^1 \right)$$

$$= \sum_{n=1}^{r} \sum_{m=1}^{r} \frac{a_m a_n}{m+n}$$

The inequality is strict unless $xp(x) \equiv 0$, that is $a_1 = a_2 = \cdots = a_r = 0$.

Prove that

$$\frac{a}{b + 2c + 3d} + \frac{b}{c + 2d + 3a} + \frac{c}{d + 2a + 3b} + \frac{d}{a + 2b + 3c} \geq \frac{2}{3}$$

for all positive real numbers a, b, c, d.

Solution

Using the Cauchy-Schwartz-Buniakowski inequality, we have

$$\left(\frac{a}{b + 2c + 3d} + \frac{b}{c + 2d + 3a} + \frac{c}{d + 2a + 3b} + \frac{d}{a + 2b + 3c} \right)$$
$$\times (a(b + 2c + 3d) + b(c + 2d + 3a) + \cdots) \geq (a + b + c + d)^2$$
$$\Rightarrow S(4)(ab + a + ad + bc + bd + cd) \geq (a + b + c + d)^2,$$

where

$$S = \left(\frac{a}{b + 2c + 3d} + \frac{b}{c + 2d + 3a} + \frac{c}{d + 2a + 3b} + \frac{d}{a + 2b + 3c} \right).$$

Now $a^2 + b^2 \geq 2ab$ from the AM–GM inequality. Likewise for the other five pairs we have the same inequality. Adding all six pairs gives

$$3(a^2 + b^2 + c^2 + d^2) \geq 2(ab + ac + \cdots)$$
$$\Rightarrow 3(a + b + c + d)^2 \geq 8(ab + ac + \cdots).$$

Therefore

$$S \geq \frac{(a + b + c + d)^2}{4(ab + ac + \cdots)} \geq \frac{(\frac{8}{3}(ab + ac + \cdots))}{4(ab + ac + \cdots)} \geq \frac{2}{3}.$$

as required.

Determine all real numbers x, y, z greater than 1, satisfying the equation

$$x + y + z + \frac{3}{x-1} + \frac{3}{y-1} + \frac{3}{z-1} = 2\left(\sqrt{x+2} + \sqrt{y+2} + \sqrt{z+2}\right).$$

Solution

For $a > 1$, $a \in \mathbb{R}$ we have the Arithmetic Mean–Geometric Mean inequality

$$a - 1 + \frac{a+2}{a-1} \geq 2\sqrt{a+2},$$

with equality if and only if $a - 1 = \frac{a+2}{a-1}$, or $a^2 - 3a - 1 = 0$, giving $a = (3 + \sqrt{13})/2$, since $a > 1$.

For each of $a = x, y, z$ we have this same result and adding them gives

$$x + y + z - 3 + \frac{x+2}{x-1} + \frac{y+2}{y-1} + \frac{z+2}{z-1} \geq 2(\sqrt{x+2} + \sqrt{y+2} + \sqrt{z+2})$$

whence

$$x + y + z + \frac{3}{x-1} + \frac{3}{y-1} + \frac{3}{z-1} \geq 2(\sqrt{x+2} + \sqrt{y+2} + \sqrt{z+2})$$

with equality if and only if $x = y = z = (3+\sqrt{13})/2$. Since there is equality, the unique solution is $x = y = z = (3 + \sqrt{13})/2$.

Find all integer solutions of

$$\frac{1}{m} + \frac{1}{n} - \frac{1}{mn^2} = \frac{3}{4}.$$

Solution

$\frac{1}{m} + \frac{1}{n} - \frac{1}{mn^2} = \frac{3}{4}$. Note $m, n \neq 0$. Then

$$\frac{n^2 + mn - 1}{mn^2} = \frac{3}{4}$$

giving

$$m = \frac{4(n^2 - 1)}{n(3n - 4)} = \frac{4(n + 1)(n - 1)}{n(3n - 4)}.$$

Now $(n + 1, n) = 1 = (n - 1, n)$, i.e. n is relatively prime to both $n + 1$ and $n - 1$. Since m is an integer, $n \mid 4(n + 1)(n - 1)$, giving $n \mid 4$. Thus $n = \pm 1, \pm 2, \pm 4$.

For

$$n = \pm 1, \quad m = 0 \text{ which is impossible}$$

$$n = -2, \quad m = \frac{4(-1)(-3)}{(-2)(-10)}, \text{ not an integer}$$

$$n = 2, \quad m = 3$$

$$n = -4, \quad m = \frac{4(-3)(-5)}{(-4)(-16)}, \text{ not an integer}$$

$$n = 4, \quad m = \frac{4(5)(3)}{4 \cdot 8}, \text{ not an integer.}$$

Thus the only solution is $(n, m) = (2, 3)$.

Let a, b, c, be rational and one of the roots of $ax^3 + bx + c = 0$ be equal to the product of the other two roots. Prove that this root is rational.

Solution

Let r_1, r_2, r_3, be the roots of the given cubic equation and also let $r_1 = r_2 r_3$. Then

$$r_1 + r_2 + r_3 = 0, \tag{1}$$

$$r_1(r_2 + r_3) + r_1 = r_1 r_2 + r_1 r_3 + r_2 r_3 = \frac{b}{a}, \tag{2}$$

$$r_1 r_2 r_3 = r_1^2 = -\frac{c}{a}. \tag{3}$$

(Note that $a \neq 0$ for the existence of three solutions.) From (1) $r_2 + r_3 = -r_1$ and substituting for $r_2 + r_3$ in (2) we find that

$$-r_1^2 + r_1 = \frac{b}{a},$$

or equivalently

$$r_1 = \frac{b}{a} + r_1^2.$$

Finally, substituting $\frac{-c}{a}$ for r_1^2 from (3), we obtain $r_1 = \frac{b-c}{a}$, which is rational.

If α, β, γ are the roots of $x^3 - x - 1 = 0$, compute

$$\frac{1+\alpha}{1-\alpha} + \frac{1+\beta}{1-\beta} + \frac{1+\gamma}{1-\gamma}.$$

Solution

If $f(x) = x^3 - x - 1 = (x - \alpha)(x - \beta)(x - \gamma)$ has roots α, β, γ standard results about roots of polynomials give $\alpha + \beta + \gamma = 0$, $\alpha\beta + \alpha\gamma + \beta\gamma = -1$, and $\alpha\beta\gamma = 1$.

Then

$$S = \frac{1+\alpha}{1-\alpha} + \frac{1+\beta}{1-\beta} + \frac{1+\gamma}{1-\gamma} = \frac{N}{(1-\alpha)(1-\beta)(1-\gamma)}$$

where the numerator simplifies to

$$\begin{aligned} N &= 3 - (\alpha + \beta + \gamma) - (\alpha\beta + \alpha\gamma + \beta\gamma) + 3\alpha\beta\gamma \\ &= 3 - (0) - (-1) + 3(1) \\ &= 7. \end{aligned}$$

The denominator is $f(1) = -1$ so the required sum is -7.

Find all real solutions to the following system of equations. Carefully justify your answer.

$$\begin{cases} \dfrac{4x^2}{1+4x^2} = y \\[2mm] \dfrac{4y^2}{1+4y^2} = z \\[2mm] \dfrac{4z^2}{1+4z^2} = x \end{cases}$$

Solution

For any t, $0 \le 4t^2 < 1 + 4t^2$, so $0 \le \dfrac{4t^2}{1+4t^2} < 1$. Thus x, y and z must be non-negative and less than 1.

Observe that if one of x y or z is 0, then $x = y = z = 0$.

If two of the variables are equal, say $x = y$, then the first equation becomes

$$\frac{4x^2}{1+4x^2} = x.$$

This has the solution $x = 0$, which gives $x = y = z = 0$ and $x = \dfrac{1}{2}$ which gives $x = y = z = \dfrac{1}{2}$.

Finally, assume that x, y and z are non-zero and distinct. Without loss of generality we may assume that either $0 < x < y < z < 1$ or $0 < x < z < y < 1$. The two proofs are similar, so we do only the first case.

We will need the fact that $f(t) = \dfrac{4t^2}{1+4t^2}$ is increasing on the interval $(0, 1)$.

To prove this, if $0 < s < t < 1$ then

$$\begin{aligned} f(t) - f(s) &= \frac{4t^2}{1+4t^2} - \frac{4s^2}{1+4s^2} \\ &= \frac{4t^2 - 4s^2}{(1+4s^2)(1+4t^2)} \\ &> 0. \end{aligned}$$

So $0 < x < y < z \Rightarrow f(x) = y < f(y) = z < f(z) = x$, a contradiction.

Hence $x = y = z = 0$ and $x = y = z = \dfrac{1}{2}$ are the only real solutions.

Given the sequence $a_0 = 1, a_1 = 2, a_{n+1} = a_n + \frac{a_{n-1}}{1+(a_{n-1})^2}, n > 1$, show that $52 < a_{1371} < 65$.

Solution

Note first that "$n > 1$" in the statement should have been "$n \geq 1$" for the problem to be correct. We show that in general,

$$\sqrt{2n+1} \leq a_n \leq \sqrt{3n+2} \text{ for all } n \geq 0. \tag{1}$$

Since when $n = 1371$, $\sqrt{2n+1} = \sqrt{2743} \approx 52.374$ and $\sqrt{3n+2} = \sqrt{4115} \approx 64.148$, $52 < a_{1371} < 65$ would follow. To establish (1), we first show by induction that

$$a_n = a_{n-1} + \frac{1}{a_{n-1}} \text{ for all } n \geq 1. \tag{2}$$

This is clearly true for $n = 1$ since $a_1 = 2 = a_0 + \frac{1}{a_0}$. Suppose (2) holds for some $n \geq 1$. Then

$$a_n = \frac{(a_{n-1})^2 + 1}{a_{n-1}} \Rightarrow \frac{1}{a_n} = \frac{a_{n-1}}{1 + (a_{n-1})^2}$$

and thus, from the given recurrence relation, we get $a_{n+1} = a_n + \frac{1}{a_n}$, completing the induction. Since clearly $a_n > 0$ for all n, we see from (2) that the sequence $\{a_n\}$ is strictly increasing. In particular, $\frac{1}{a_{n-1}^2} \leq 1$ for all $n \geq 1$ and so from $a_n^2 = a_{n-1}^2 + 2 + \frac{1}{a_{n-1}^2}$ we get

$$a_{n-1}^2 + 2 < a_n^2 \leq a_{n-1}^2 + 3 \text{ for all } n \geq 1. \tag{3}$$

Now we use (3) and induction to establish (1). The case when $n = 0$ is trivial since $a_0 = 1 < \sqrt{2}$. Suppose (1) holds for some $n \geq 0$. Then by (3),

$$a_{n+1} \leq \sqrt{a_n^2 + 3} \leq \sqrt{3n+2+3} = \sqrt{3(n+1)+2}$$

and

$$a_{n+1} > \sqrt{a_n^2 + 2} \geq \sqrt{2n+1+2} = \sqrt{2(n+1)+1}$$

and our proof is complete.

Let $\{b_n\}$ be sequence of positive real numbers such that

$$\text{for each } n \geq 1, \quad b_{n+1}^2 \geq \frac{b_1^2}{1^3} + \frac{b_2^2}{2^3} + \cdots + \frac{b_n^2}{n^3}.$$

Show that there is a natural number K such that

$$\sum_{n=1}^{K} \frac{b_{n+1}}{b_1 + b_2 + \cdots + b_n} > \frac{1993}{1000}.$$

Solution

$$(1^3 + 2^3 + \cdots + n^3)(b_{n+1}^2)$$
$$\geq (1^3 + 2^3 + \cdots + n^3)\left(\frac{b_1^2}{1^3} + \frac{b_2^2}{2^3} + \cdots + \frac{b_n^2}{n^3}\right)$$
$$\geq (b_1 + b_2 + \cdots + b_n)^2$$

by the Cauchy–Schwartz inequality. Thus

$$\frac{b_{n+1}^2}{(b_1 + b_2 + \cdots + b_n)^2} \geq \frac{1}{1^3 + 2^3 + \cdots + n^3} = \left[\frac{2}{n(n+1)}\right]^2.$$

It follows that

$$\frac{b_{n+1}}{b_1 + b_2 + \cdots + b_n} \geq \frac{2}{n(n+1)}$$

since the sequence $\{b_n\}$ has only positive real terms. Thus

$$\sum_{n=1}^{K} \frac{b_{n+1}}{b_1 + b_2 + \cdots + b_n} \geq \sum_{n=1}^{K} \frac{2}{n(n+1)}$$
$$= 2\sum_{n=1}^{K} \frac{1}{n(n+1)}$$
$$= 2\sum_{n=1}^{K} \left(\frac{1}{n} - \frac{1}{n+1}\right)$$
$$= 2\left(1 - \frac{1}{K+1}\right)$$
$$= \frac{2K}{K+1}.$$

By setting $K = 999$ we have

$$\sum_{n=1}^{999} \frac{b_{n+1}}{b_1 + \cdots + b_n} \geq \frac{2(999)}{999 + 1} = \frac{1998}{1000} > \frac{1993}{1000}.$$

as required.

The real numbers α, β satisfy the equations

$$\alpha^3 - 3\alpha^2 + 5\alpha - 17 = 0, \qquad \beta^3 - 3\beta^2 + 5\beta + 11 = 0.$$

Find $\alpha + \beta$.

Solution

Define $f(x) = x^3 - 3x^2 + 5x$. We show that if $f(\alpha) + f(\beta) = 6$, then $\alpha + \beta = 2$. Since $f(x) = (x-1)^3 + 2(x-1) + 3$, we have

$$f(\alpha) - 3 = (\alpha - 1)^3 + 2(\alpha - 1)$$
$$f(\beta) - 3 = (\beta - 1)^3 + 2(\beta - 1)$$

Adding gives

$$0 = (\alpha - 1)^3 + (\beta - 1)^3 + 2(\alpha + \beta - 2)$$
$$= (\alpha + \beta - 2)[(\alpha - 1)^2 + (\alpha - 1)(\beta - 1) + (\beta - 1)^2 + 2]$$

and, since the second factor is positive, we obtain the result.

For non-negative integers n, r the binomial coefficient $\binom{n}{r}$ denotes the number of combinations of n objects chosen r at a time, with the convention that $\binom{n}{0} = 1$ and $\binom{n}{r} = 0$ if $n < r$. Prove the identity

$$\sum_{d=1}^{\infty} \binom{n-r+1}{d}\binom{r-1}{d-1} = \binom{n}{r}$$

for all integers n, r with $1 \le r \le n$.

Solution

We use a combinatorial argument to establish the obviously equivalent identity

$$\sum_{d=1}^{k} \binom{n-r+1}{d}\binom{r-1}{r-d} = \binom{n}{r} \qquad (*)$$

where $k = \min\{r, n - r + 1\}$. It clearly suffices to demonstrate that the left hand side of $(*)$ counts the number of ways of selecting r objects from n distinct objects (without replacements). Let $|S_2| = r - 1$. For each fixed $d = 1, 2, \ldots, k$, any selection of d objects from S_1 ($S \backslash S_2$) together with any selection of $r - d$ objections from S_2 would yield a selection of r objects from S. The total number of such selections is $\binom{n-r+1}{d}\binom{r-1}{r-d}$. Conversely, each selection of r objects from S clearly must arise in this manner. Summing over $d = 1, 2, \ldots$, $(*)$ follows.

For every positive integer n, $n?$ is defined as follows:

$$n? = \begin{cases} 1 & for\ n = 1 \\ \frac{n}{(n-1)?} & for\ n \geq 2 \end{cases}$$

Prove $\sqrt{1992} < 1992? < \frac{4}{3}\sqrt{1992}$.

Solution

Using the more convenient notation $f(n)$ for $n?$, we show that in general

$$\sqrt{n+1} < f(n) < \frac{4}{3}\sqrt{n} \qquad (*)$$

for all **even** $n \geq 6$. In particular, for $n = 1992$, we would get $\sqrt{1993} < f(1992) < \frac{4}{3}\sqrt{1992}$.

First note that $f(n) = \frac{n}{f(n-1)} = \frac{n}{n-1}f(n-2)$ for all $n \geq 3$. If $N = 2k$ where $k \geq 2$, then multiplying $f(2q) = \frac{2q}{2q-1}f(2q-2)$ for $q = 2, 3, \ldots, k$, we get

$$
\begin{aligned}
f(2k) &= \frac{4}{3} \cdot \frac{6}{5} \cdots \frac{2k}{2k-1} \cdot f(2) \\
&= \left(\frac{2}{1}\right) \cdot \left(\frac{4}{3}\right) \cdot \left(\frac{6}{5}\right) \cdots \left(\frac{2k}{2k-1}\right) > \left(\frac{3}{2}\right)\left(\frac{5}{4}\right)\left(\frac{7}{6}\right) \cdots \left(\frac{2k+1}{2k}\right).
\end{aligned}
$$

Hence

$$(f(2k))^2 > \frac{2 \cdot 4 \cdot 6 \cdots 2k}{1 \cdot 3 \cdot 5 \cdots (2k-1)} \cdot \frac{3 \cdot 5 \cdot 7 \cdots (2k+1)}{2 \cdot 4 \cdot 6 \cdots 2k} = 2k + 1,$$

from which it follows that

$$f(n) = f(2k) > \sqrt{2k+1} = \sqrt{n+1}. \qquad (1)$$

On the other hand, for $k \geq 3$ we have

$$
\begin{aligned}
2(2k) &= \left(\frac{2}{3}\right)\left(\frac{4}{5}\right)\left(\frac{6}{7}\right) \cdots \left(\frac{2k-2}{2k-1}\right) \cdot 2k \\
&< \left(\frac{2}{3}\right)\left(\frac{5}{6}\right)\left(\frac{7}{8}\right) \cdots \left(\frac{2k-1}{2k}\right) 2k.
\end{aligned}
$$

Hence

$$
\begin{aligned}
(f(2k))^2 &< \left(\frac{2}{3}\right)^2 \cdot \frac{4 \cdot 6 \cdots (2k-2)}{5 \cdot 7 \cdots (2k-1)} \cdot \frac{5 \cdot 7 \cdots (2k-1)}{6 \cdot 8 \cdots 2k} \cdot (2k)^2 \\
&= \left(\frac{2}{3}\right)^2 \cdot 4 \cdot 2k,
\end{aligned}
$$

from which it follows that

$$f(n) = f(2k) < \frac{4}{3}\sqrt{2k} = \frac{4}{3}\sqrt{n}. \qquad (2)$$

The result follows from (1) and (2).

Determine all natural numbers $x, y \geq 1$ such that $2^x - 3^y = 7$.

Solution

First of all we note that both numbers x and y must be even. Suppose to the contrary that one of the numbers is odd.

If x is odd, the number $2^x + 1$ (which we have from the factorization $A^x + 1 = (A+1)(A^{x-1} - A^{x-2} + \cdots - A + 1)$ that $2^x + 1$) is a multiple of 3. Consequently $2^x - 3^y \equiv -1 \bmod 3$, while $7 \equiv 1 \bmod 3$, and the given equation is invalid mod 3.

If y is odd, we will use the modular technique as in the previous case, but this time modulo 8. We have $3^2 \equiv 1 \bmod 8$. It follows that $3^{2k+1} \equiv 3 \bmod 8$ for $k = 0, 1, 2, \ldots$.

Consequently $3^y + 7 \equiv 0 \bmod 8$ and since $2^x \equiv 3^y + 7$ we must have $x \leq 2$. If $x = 1$, we have $2 - 3^y = 7$, which is impossible. If $x = 2$, we have $2^2 - 3^y = 7$, which is also impossible.

Suppose that the numbers x and y are even. So $x = 2l$, $y = 2m$, with l and m natural numbers. The given equation can then be written in the form $(2^l + 3^m)(2^l - 3^m) = 7$, where $2^l + 3^m$ and $2^l - 3^m$ are natural numbers, which implies that $2^l + 3^m = 7$ and $2^l - 3^m = 1$. These two equations determine the values of l, m, namely $l = 2$, $m = 1$ for which we have $x = 4$, $y = 2$.

The only two natural numbers $x, y \geq 1$ such that $2^x - 3^y = 7$ are therefore $x = 4$ and $y = 2$.

Determine all real solutions x, y, z of the system of equations:

$$
\begin{aligned}
x^3 + y &= 3x + 4, \\
2y^3 + z &= 6y + 6, \\
3z^3 + x &= 9z + 8.
\end{aligned}
$$

Solution

From $x^3 + y = 3x + 4$, we have

$$x^3 - 1 - 1 - 3x = 2 - y,$$

or

$$(x - 2)(x + 1)^2 = 2 - y. \tag{1}$$

From $2y^3 - 2 - 2 - 6y = 2 - z$, we have

$$2(y - 2)(y + 1)^2 = (2 - z). \tag{2}$$

and $3z^3 - 3 - 3 - 9z = 2 - x$ gives

$$3(z - 2)(z + 1)^2 = (2 - x). \tag{3}$$

so that

$$
\begin{aligned}
(x - 2)(x + 1)^2 &= -(y - 2), \\
2(y - 2)(y + 1)^2 &= -(z - 2), \\
3(z - 2)(z + 1)^2 &= -(x - 2),
\end{aligned}
$$

and

$$(x - 2)(y - 2)(z - 2)\left((x + 1)^2(y + 1)^2(z + 1)^2 + \frac{1}{6}\right) = 0.$$

As the last factor is always positive for real x, y, z, we have

$$(x - 2)(y - 2)(z - 2) = 0.$$

This gives at least one of $x = 2$, $y = 2$, $z = 2$. In conjunction with (1), (2) and (3) this gives the unique solution $x = y = z = 2$.

Show: For all real numbers $a, b \geq 0$ the following chain of inequalities is valid

$$\left(\frac{\sqrt{a} + \sqrt{b}}{2}\right)^2 \leq \frac{a + \sqrt[3]{a^2 b} + \sqrt[3]{ab^2} + b}{4}$$

$$\leq \frac{a + \sqrt{ab} + b}{3} \leq \sqrt{\left(\frac{\sqrt[3]{a^2} + \sqrt[3]{b^2}}{2}\right)^3}.$$

Also, for all three inequalities determine the cases of equality.

Solution

1. $\left(\dfrac{\sqrt{a} + \sqrt{b}}{2}\right)^2 \leq \dfrac{\sqrt[3]{a^2}(\sqrt[3]{a} + \sqrt[3]{b}) + \sqrt[3]{b^2}(\sqrt[3]{a} + \sqrt[3]{b})}{4}$

is equivalent to

$$(\sqrt{a} + \sqrt{b})^2 \leq (\sqrt[3]{a^2} + \sqrt[3]{b^2})(\sqrt[3]{a} + \sqrt[3]{b}),$$

which holds by the Cauchy inequality.

Let $\sqrt[6]{a} = A$, $\sqrt[6]{b} = B$, $(A^3 + B^3)^2 \leq (A^4 + B^4)(A^2 + B^2)$.

2. $3(a + b) + 3\sqrt[3]{ab}(\sqrt[3]{a} + \sqrt[3]{b}) \leq 4(a + \sqrt{ab} + b)$,

or equivalently

$$a + 3\sqrt[3]{a^2 b} + 3\sqrt[3]{ab^2} + b \leq 2(a + 2\sqrt{ab} + b),$$

or

$$(\sqrt[3]{a} + \sqrt[3]{b})^3 \leq 2(\sqrt{a} + \sqrt{b})^2.$$

or

$$\left(\frac{A^2 + B^2}{2}\right)^3 \leq \left(\frac{A^3 + B^3}{2}\right)^2,$$

(with A, B as above), a known inequality.

3. $\dfrac{a + \sqrt{ab} + b}{3} \leq \sqrt{\left(\dfrac{\sqrt[3]{a^2} + \sqrt[3]{b^2}}{2}\right)^3}.$

With A and B as above this is equivalent to

$$\left(\frac{A^6 + A^3 B^3 + B^6}{3}\right)^2 \leq \left(\frac{A^4 + B^4}{2}\right)^3.$$

For this it is enough to prove that

$$\left(\frac{A^4 + B^4}{2}\right)^3 - \left(\frac{A^6 + A^3 B^3 + B^6}{3}\right)^2 \geq 0.$$

or

$$9(A^4 + B^4)^3 - 8(A^6 + A^3 B^3 + B^6)^2$$
$$= A^{12} - 16A^9 B^3 + 27A^8 B^4 - 24A^6 B^6 + 27A^4 B^8 - 16A^3 B^9 + B^{12}$$
$$= (A - B)^4 [A^8 + 4A^7 B + 10A^6 B^2 + 4A^5 B^3 - 2A^4 B^4$$
$$+ 4A^3 B^5 + 10A^2 B^6 + 4AB^7 + B^8]$$
$$\geq (A - B)^4 (A^3 B^3 (A - B)^2) \geq 0.$$

There are real numbers a, b, c, such that $a \geq b \geq c > 0$. Prove that

$$\frac{a^2 - b^2}{c} + \frac{c^2 - b^2}{a} + \frac{a^2 - c^2}{b} \geq 3a - 4b + c.$$

Solution

From $a \geq b \geq c > 0$ we have

$$\frac{a + b}{c} \geq 2, \quad 0 < \frac{b + c}{a} \leq 2 \quad \text{and} \quad \frac{a + c}{b} \geq 1.$$

Now we get

$$\frac{a^2 - b^2}{c} \geq 2(a - b), \quad \text{because } a \geq b;$$

$$\frac{c^2 - b^2}{a} \geq 2(c - b), \quad \text{because } c \leq b;$$

and

$$\frac{a^2 - c^2}{b} \geq a - c, \quad \text{because } a \geq c.$$

After addition of these inequalities, we have

$$\frac{a^2 - b^2}{c} + \frac{c^2 - b^2}{a} + \frac{a^2 - c^2}{b} \geq 2(a - b) + 2(c - b) + (a - c),$$

that is,

$$\frac{a^2 - b^2}{c} + \frac{c^2 - b^2}{a} + \frac{a^2 - c^2}{b} \geq 3a - 4b + c.$$

The equality holds if and only if $a = b = c > 0$.

Prove that there are no real numbers x, y, z, such that

$$x^2 + 4yz + 2z = 0,$$

$$x + 2xy + 2z^2 = 0,$$

$$2xz + y^2 + y + 1 = 0.$$

Solution

Label the three given equations (1), (2) and (3), (in that order), respectively. If $x = 0$ or $z = 0$, then from (3), $y^2 + y + 1 = 0$, which has no real solutions. Hence we may assume that $xz \neq 0$. From (1) and (2), we get $x^2 = -2z(2y + 1)$ and $2z^2 = -x(2y + 1)$, which, when multiplied gives $2x^2z^2 = 2xz(2y + 1)^2$ or $xz = (2y + 1)^2$. Substituting into (3) we get

$$2(2y + 1)^2 + y^2 + y + 1 = 0$$

or

$$3y^2 + 3y + 1 = 0,$$

which has no solution.

Let a, b, and c be positive real numbers such that $abc = 1$. Prove that

$$\frac{1}{a^3(b+c)} + \frac{1}{b^3(c+a)} + \frac{1}{c^3(a+b)} \geq \frac{3}{2}.$$

Solution

By the Cauchy–Schwartz inequality

$$[a(b+c) + b(c+a) + c(a+b)]\left[\frac{1}{a^3(b+c)} + \frac{1}{b^3(c+a)} + \frac{1}{c^3(a+b)}\right]$$

$$\geq \left(\frac{1}{a} + \frac{1}{b} + \frac{1}{c}\right)^2.$$

or

$$2(ab + ac + bc)\left[\frac{1}{a^3(b+c)} + \frac{1}{b^3(c+a)} + \frac{1}{c^3(a+b)}\right]$$

$$\geq \frac{(ab + ac + bc)^2}{(abc)^2},$$

or

$$\frac{1}{a^3(b+c)} + \frac{1}{b^3(c+a)} + \frac{1}{c^3(a+b)} \geq \frac{ab + ac + bc}{2}.$$

because $abc = 1$.

Also

$$\frac{ab + ac + bc}{3} \geq \sqrt[3]{a^2b^2c^2} = 1.$$

Therefore

$$\frac{1}{a^3(b+c)} + \frac{1}{b^3(c+a)} + \frac{1}{c^3(a+b)} \geq \frac{3}{2}$$

holds.

Prove that

$$\frac{1}{1999} < \frac{1}{2} \cdot \frac{3}{4} \cdot \frac{5}{6} \cdots \frac{1997}{1998} < \frac{1}{44}.$$

Solution

Let $P = \frac{1}{2} \cdot \frac{3}{4} \cdot \ldots \cdot \frac{1997}{1998}$. Then $\frac{1}{2} > \frac{1}{3}$ [because $2 < 3$], $\frac{3}{4} > \frac{3}{5}$ [because $4 < 5$], $\ldots \ldots \frac{1997}{1998} > \frac{1997}{1999}$ [because $1998 < 1999$].

So

$$P > \frac{1}{3} \cdot \frac{3}{5} \cdot \ldots \cdot \frac{1997}{1999} = \frac{1}{1999}. \tag{1}$$

Also $\frac{1}{2} < \frac{2}{3}$ [because $1 \cdot 3 < 2 \cdot 2$], $\frac{3}{4} < \frac{4}{5}$ [because $3 \cdot 5 < 4 \cdot 4$], $\ldots$, $\frac{1997}{1998} < \frac{1998}{1999}$ [because $1997 \cdot 1999 = 1998^2 - 1 < 1998^2$].

So $P < \frac{2}{3} \cdot \frac{4}{5} \cdot \ldots \cdot \frac{1998}{1999} = \underbrace{\left(\frac{2}{1} \cdot \frac{4}{3} \cdot \frac{6}{5} \cdot \ldots \cdot \frac{1998}{1997} \right)}_{1/P} \frac{1}{1999}.$

Hence $P^2 < \frac{1}{1999} < \frac{1}{1936} = \frac{1}{44^2}$ and $P < \frac{1}{44}$. $\tag{2}$

Then (1) and (2) give $\frac{1}{1999} < P < \frac{1}{44}$.

Let n be a positive integer, $a_1, a_2, \ldots, a_n$ positive real numbers and $s = a_1 + a_2 + \cdots + a_n$. Prove that

$$\sum_{i=1}^{n} \frac{a_i}{s - a_i} \geq \frac{n}{n-1} \qquad \text{and} \qquad \sum_{i=1}^{n} \frac{s - a_i}{a_i} \geq n(n-1).$$

Solution

$$\sum_{i=1}^{n} \frac{s - a_i}{a_i} = \sum_{i=1}^{n} \left(\frac{s}{a_i} - 1 \right) = \sum_{i=1}^{n} \frac{s}{a_i} - n$$

$$\sum_{i=1}^{n} \frac{a_i}{s} = 1 \qquad \text{and} \qquad \sum_{i=1}^{n} \frac{s}{a_i} \sum_{i=1}^{n} \frac{a_i}{s} \geq \left(\sum 1 \right)^2 = n^2,$$

by the Cauchy–Schwarz inequality.

Thus

$$\sum_{i=1}^{n} \frac{s}{a_i} \geq n^2.$$

Hence $\sum_{i=1}^{n} \dfrac{s - a_i}{a_i} \geq n^2 - n = n(n-1)$.

To prove the first inequality, first note that

$$\sum_{i=1}^{n} 1 \sum_{i=1}^{n} a_i^2 \geq \left(\sum_{i=1}^{n} a_i \right)^2 = s^2.$$

Hence $\sum_{i=1}^{n} a_i^2 \geq \dfrac{s^2}{n}$.

By the Cauchy–Schwarz inequality,

$$\sum_{i=1}^{n} a_i(s - a_i) \sum_{i=1}^{n} \frac{a_i}{s - a_i} \geq \left(\sum_{i=1}^{n} a_i \right)^2 = s^2.$$

Therefore

$$\sum_{i=1}^{n} \frac{a_i}{s - a_i} \geq \frac{s^2}{\sum_{i=1}^{n} a_i(s - a_i)} = \frac{s^2}{s \sum_{i=1}^{n} a_i - \sum_{i=1}^{n} a_i^2}$$

$$\geq \frac{s^2}{s^2 - \frac{s^2}{n}} = \frac{1}{1 - \frac{1}{n}} = \frac{n}{n-1}.$$

So, both inequalities are proved.

Show that if x, y, z are positive integers such that $x^2 + y^2 + z^2 = 1993$, then $x + y + z$ is not a perfect square.

Solution

We show that the result holds for *nonnegative* integers x, y, and z. Without loss of generality, we may assume that $0 \leq x \leq y \leq z$. Then

$$3z^2 \geq x^2 + y^2 + z^2 = 1993$$

implies that

$$z^2 \geq 665, \quad z \geq 26.$$

On the other hand $z^2 \leq 1993$ implies that $z \leq 44$ and thus $26 \leq z \leq 44$.

Suppose that $x + y + z = k^2$ for some nonnegative integer k. By the Cauchy–Schwarz Inequality we have

$$k^4 = (x + y + z)^2 \leq (1^2 + 1^2 + 1^2)(x^2 + y^2 + z^2) = 5979$$

and so $k \leq \lfloor \sqrt[4]{5979} \rfloor = 8$. Since $k^2 \geq z \geq 26$, $k \geq 6$. Furthermore, since $x^2 + y^2 + z^2$ is odd, it is easily seen that $x + y + z$ must be odd, which implies that k is odd. Thus $k = 7$ and we have $x + y + z = 49$.

Let $z = 26 + d$, where $0 \leq d \leq 18$. Then

$$x + y = 23 - d \Rightarrow y \leq 23 - d \Rightarrow x^2 + y^2 \leq 2y^2$$
$$\leq 2(23 - d)^2 = 1058 - 92d + 2d^2. \qquad (1)$$

On the other hand, from $x^2 + y^2 + z^2 = 1993$ we get

$$x^2 + y^2 = 1993 - z^2 = 1993 - (26 + d)^2 = 1317 - 52d - d^2. \qquad (2)$$

From (1) and (2), we get

$$1317 - 52d - d^2 \leq 1058 - 92d + 2d^2$$

or

$$3d^2 - 40d \geq 259$$

which is clearly impossible since $3d^2 - 40d = d(3d - 40) \leq 18 \times 14 = 252$. This completes the proof.

Given that $a^2 + b^2 + (a + b)^2 = c^2 + d^2 + (c + d)^2$, prove that
$a^4 + b^4 + (a + b)^4 = c^4 + d^4 + (c + d)^4$.

Solution

This follows immediately from the fact that

$$
\begin{aligned}
x^4 + y^4 + (x + y)^4 &= 2\left(x^4 + 2x^3y + 3x^2y^2 + 2xy^3 + y^4\right) \\
&= 2\left(x^2 + xy + y^2\right)^2 = \frac{1}{2}\left[x^2 + y^2 + (x + y)^2\right]^2
\end{aligned}
$$

for all x, y.

Prove that the product of the 99 numbers $\frac{k^3-1}{k^3+1}$, $k = 2, 3, \ldots, 100$, is greater than $\frac{2}{3}$.

Solution

Let $P(n) = \prod_{k=2}^{n} \frac{k^3-1}{k^3+1}$ where $n \geq 2$. We show that in general

$$P(n) = \frac{2(n^2 + n + 1)}{3n(n + 1)}$$

from which it follows that $P(n) > \frac{2}{3}$. Since

$$(k + 1)^2 - (k + 1) + 1 = k^2 + k + 1,$$

we have

$$
\begin{aligned}
P(n) &= \prod_{k=2}^{n} \frac{(k - 1)(k^2 + k + 1)}{(k + 1)(k^2 - k + 1)} \\
&= \frac{\prod_{k=0}^{n-2}(k + 1)}{\prod_{k=2}^{n}(k + 1)} \cdot \frac{\prod_{k=2}^{n}(k^2 + k + 1)}{\prod_{k=1}^{n-1} k^2 + k + 1} \\
&= \frac{1 \cdot 2}{n(n + 1)} \cdot \frac{n^2 + n + 1}{3} = \frac{2(n^2 + n + 1)}{3n(n + 1)}.
\end{aligned}
$$

Remark: It is easy to see that the sequence $\{P(n)\}$, $n = 2, 3, \ldots$ is strictly decreasing and from the result established above we see that $\lim_{n \to \infty} P(n) = \frac{2}{3}$.

Find all integers satisfying the equation

$$2^x \cdot (4 - x) = 2x + 4.$$

Solution

$$2^x = \frac{2(x + 2)}{4 - x} > 0, \quad \text{so} \quad -2 < x < 4.$$

Checking

$$x = -1 \quad \frac{1}{2} \neq \frac{2(-1) + 4}{4 - (-1)} = \frac{2}{5}$$

$$x = 0 \quad 1 = \frac{2(0 + 2)}{4 - 0} \quad \text{is a solution}$$

$$x = 1 \quad 2 = \frac{2(1 + 2)}{4 - 1} \quad \text{is a solution}$$

$$x = 2 \quad 4 = \frac{2(2 + 2)}{4 - 2} \quad \text{is a solution}$$

$$x = 3 \quad 8 = \frac{2(3 + 2)}{4 - 3} \quad \text{is not a solution}$$

Prove that for any positive $x_1, x_2, \ldots, x_n, y_1, y_2, \ldots, y_n$ the inequality

$$\sum_{i=1}^{n} \frac{1}{x_i y_i} \geq \frac{4n^2}{\sum_{i=1}^{n}(x_i + y_i)^2}$$

holds.

Solution

Now,

$$\frac{1}{xy} \geq \frac{4}{(x+y)^2}$$

since $(x+y)^2 \geq 4xy$, as $(x-y)^2 \geq 0$. So

$$\sum_{i=1}^{n} \frac{1}{x_i y_i} \geq \sum_{i=1}^{n} \frac{4}{(x_i + y_i)^2} \qquad (*)$$

Lemma. $(a_1 + a_2 + \cdots + a_n)(a_2 a_3 \cdots a_n + a_1 a_3 a_4 \cdots a_n + a_1 a_2 a_4 \cdots a_n + \cdots + a_1 a_2 \cdots a_{n-1}) \geq n^2 a_1 a_2 \cdots a_n$.

This follows from separate applications of the AM–GM inequality to the two terms on the left. It follows that

$$a_1 + a_2 + \cdots + a_n \geq \frac{n^2}{\frac{1}{a_1} + \frac{1}{a_2} + \cdots + \frac{1}{a_n}}.$$

Now put

$$a_i = \frac{1}{(x_i + y_i)^2}, \qquad i = 1, \ldots, n$$

and then

$$\frac{1}{(x_1 + y_1)^2} + \frac{1}{(x_2 + y_2)^2} + \cdots + \frac{1}{(x_n + y_n)^2} \geq \frac{n^2}{\sum_{i=1}^{n}(x_i + y_i)^2}.$$

Combining this with $(*)$ shows

$$\sum_{i=1}^{n} \frac{1}{x_i y_i} \geq \frac{4n^2}{\sum_{i=1}^{n}(x_i + y_i)^2}.$$

For nonnegative integers n, r the binomial coefficient $\binom{n}{r}$ denotes the number of combinations of n objects chosen r at a time, with the convention that $\binom{n}{0} = 1$ and $\binom{n}{r} = 0$ if $n < r$. Prove the identity

$$\sum_{d=1}^{\infty} \binom{n-r+1}{d}\binom{r-1}{d-1} = \binom{n}{r}$$

for all integers n, r with $1 \le r \le n$.

Solution

We have

$$
\begin{aligned}
(1+x)^n &= (1+x)^{n-r+1}(1+x)^{r-1} \\
&= \left[\sum_{i=0}^{m-r+1}\binom{n-r+1}{i}x^i\right]\left[\sum_{j=0}^{r-1}\binom{r-1}{j}x^j\right] \\
&= \sum_{i=0}^{m-r+1}\sum_{j=0}^{r-1}\binom{n-r+1}{i}\binom{r-1}{j}x^{i+j}
\end{aligned}
$$

The coefficient of x^r is

$$\sum_{d=0}^{r}\binom{n-r+1}{d}\binom{r-1}{r-d} = \sum_{d=1}^{r}\binom{n-r+1}{d}\binom{r-1}{d-1}$$

Let x be a real number with $0 < x < \pi$. Prove that, for all natural numbers n, the sum

$$\sin x + \frac{\sin 3x}{3} + \frac{\sin 5x}{5} + \cdots + \frac{\sin(2n-1)x}{2n-1}$$

is positive.

Solution

We know that

$$2 \sin x \sin(2k-1)x = \cos(2k-2)x - \cos 2kx.$$

Hence

$$2 \sin x \left(\sin x + \frac{\sin 3x}{3} + \cdots + \frac{\sin(2n-1)x}{2n-1} \right)$$

$$= 1 - \cos 2x + \frac{\cos 2x - \cos 4x}{3} + \cdots + \frac{\cos(2n-2)x - \cos 2nx}{2n-1}$$

$$= 1 - \cos 2x \left(1 - \frac{1}{3} \right) - \cos 4x \left(\frac{1}{3} - \frac{1}{5} \right) - \cdots - \frac{\cos 2nx}{2n-1}$$

$$\geq 1 - \left[\left(1 - \frac{1}{3} \right) + \left(\frac{1}{3} - \frac{1}{5} \right) + \cdots + \frac{1}{2n-1} \right] = 0.$$

It is given that $\cos x = \cos y$ and $\sin x = -\sin y$. Prove that $\sin 1994x + \sin 1994y = 0$.

Solution

More generally, we prove that $\sin mx + \sin my = 0$ where m is an integer.

Since $\sin mx + \sin my = 2\sin\left(m\frac{x+y}{2}\right)\cos\left(m\left(\frac{x-y}{2}\right)\right)$, it is sufficient to show that $\sin\left(\frac{x+y}{2}\right) = 0$ since $\sin\left(m\frac{x+y}{2}\right) = 0$ follows easily by induction from

$$\sin\left((m+1)\frac{x+y}{2}\right) = \sin\left(m\frac{(x+y)}{2}\right)\cos\left(\frac{x+y}{2}\right)$$
$$+ \cos\left(m\frac{(x+y)}{2}\right)\sin\left(\frac{x+y}{2}\right).$$

Now,

$$\cos x = \cos y \iff \cos x - \cos y = 0$$
$$\iff -2\sin\frac{x+y}{2}\sin\frac{x-y}{2} = 0 \qquad (1)$$

and

$$\sin x = -\sin y \iff \sin x + \sin y = 0$$
$$\iff 2\sin\frac{x+y}{2}\cos\frac{x-y}{2} = 0. \qquad (2)$$

Squaring each of (1) and (2) and adding, we find

$$4\sin^2\left(\frac{x+y}{2}\right)\underbrace{\left(\sin^2\frac{(x-y)}{2} + \cos^2\frac{(x-y)}{2}\right)}_{=1} = 0.$$

Hence $\sin\frac{x+y}{2} = 0$.

It is given that $a > 0, b > 0, c > 0, a + b + c = abc$. Prove that at least one of the numbers a, b, c exceeds $17/10$.

Solution

We show, in general, that if $x_i > 0$ for $i = 1, 2, \ldots, n$, such that $\sum_{i=1}^{n} x_i = \prod_{i=1}^{n} x_i$, then $\max\{x_i : i = 1, 2, \ldots, n\} \geq n^{1/(n-1)}$. In particular, when $n = 3$ we get

$$\max\{x_1, x_2, x_3\} \geq \sqrt{3} > 1.7 = \frac{17}{10}.$$

By the arithmetic-geometric mean inequality we have

$$\left(\frac{1}{n} \sum_{i=1}^{n} x_i \right)^n \geq \prod_{i=1}^{n} x_i = \sum_{i=1}^{n} x_i.$$

and thus $\sum_{i=1}^{n} x_i \geq n^{n/(n-1)}$.

Without loss of generality, we may assume that $\max\{x_i : i = 1, 2, \ldots, n\} = x_n$. Then $n x_n \geq \sum_{i=1}^{n} x_i \geq n^{n/(n-1)}$ from which $x_n \geq n^{1/(n-1)}$ follows.

Solve the equation $1! + 2! + 3! + \cdots + n! = m^3$ in natural numbers.

Solution

We solve the more general problem of finding all solutions of the Diophantine equation $1! + 2! + 3! + \cdots + n! = m^k$ in natural numbers, n, m, and k. For convenience, let $S_n = 1! + 2! + 3! + \cdots + n!$.

When $k = 1$ clearly $m = S_n$ is the only solution for any n.

When $k = 2$ we claim that the equation $S_n = m^2$ has exactly two solutions: $n = m = 1$ and $n = m = 3$. Note first that $d! \equiv 0 \pmod{10}$ for all $d \geq 5$ and $S_4 = 1 + 2 + 6 + 24 = 33 \equiv 3 \pmod{10}$. Hence $S_n \equiv 3 \pmod{10}$ for all $n \geq 4$. However, it is easy to see that the last digit of a perfect square can never be 3 and so there are no solutions if $n \geq 4$. Checking the cases when $n = 1, 2, 3$ directly reveals that there are precisely two solutions, as given above.

When $k \geq 3$ we show that $n = m = 1$ is the only solution. If $n \geq 2$ then clearly $S_n \equiv 0 \pmod{3}$. But $m^k \equiv 0 \pmod{3}$ implies $m \equiv 0 \pmod{3}$ and so $m^k \equiv 0 \pmod{27}$ as $k \geq 3$. Since $d! \equiv 0 \pmod{27}$ for all $d \geq 9$ and since

$$S_8 = 1 + 2 + 6 + 24 + 120 + 720 + 5040 + 40320 = 46233 \not\equiv 0 \pmod{27}$$

there are no solutions if $n \geq 8$. On the other hand, direct checking shows that for $n = 2, 3, 4, 5, 6, 7$, $S_n = 3, 9, 33, 153, 873$, and 5913, none of which is a perfect k^{th} power for any $k \geq 3$. Finally, it is trivial to see that $n = m = 1$ is a solution.

It is given that x and y are positive integers and $3x^2 + x = 4y^2 + y$.
Prove that $x - y$, $3x + 3y + 1$ and $4x + 4y + 1$ are squares of integers.

Solution

Note first that the given equation implies the following two equations:

$$(x - y)(3x + 3y + 1) = y^2 \qquad (1)$$
$$(x - y)(4x + 4y + 1) = x^2 \qquad (2)$$

(1) × (2) yields $(x - y)^2(3x + 3y + 1)(4x + 4y + 1) = (xy)^2$ which implies that $(3x + 3y + 1)(4x + 4y + 1)$ is a perfect square. But clearly $\gcd(3x + 3y + 1, 4x + 4y + 1) = 1$ since $4(3x + 3y + 1) - 3(4x + 4y + 1) = 1$. Therefore, $3x + 3y + 1$ and $4x + 4y + 1$ are both squares, which, together with (1) (or (2)), implies that $x - y$ is also a square.

It is given that $0 \leq x_i \leq 1$, $i = 1, 2, \ldots, n$. Find the maximum of the expression

$$\frac{x_1}{x_2 x_3 \ldots x_n + 1} + \frac{x_2}{x_1 x_3 x_4 \ldots x_n + 1} + \cdots + \frac{x_n}{x_1 x_2 \ldots x_{n-1} + 1}.$$

Solution

Since the expression is convex in each of the variables, the maximum value is achieved when each variable takes on 0 or 1. Clearly this occurs when one variable is 0 and the rest are 1 giving the maximum value of $n - 1$. The same maximum occurs if any of the numerators x_i are replaced by $x_i^{\alpha_i}$ where $\alpha_i \geq 1$.

A similar result, using convexity, that

$$\sum \frac{x_i^u}{1 + s - x_i} + \prod (1 - x_i)^v \leq 1,$$

where $0 \leq x_i \leq 1$, $u, v \geq 1$, $s = \sum x_i$ and the sum and product are over $i = 1, 2, \ldots, n$, is given in [1].

It is given that $a > 0$, $b > 0$, $c > 0$, $a + b + c = abc$. Prove that at least one of the numbers a, b, c exceeds $17/10$.

Solution

In any triangle ABC the following identity holds

$$\tan \alpha + \tan \beta + \tan \gamma = \tan \alpha \tan \beta \tan \gamma.$$

Hence, in this problem a, b, c can be considered to be the tangents of angles of an acute angled triangle. At least one of the angles is at least $\frac{\pi}{3}$, and $\tan \left(\frac{\pi}{3} \right) = \sqrt{3} > \frac{17}{10}$.

Second solution.
We may suppose $a \geq b \geq c$,

$$abc = a + b + c \geq 3c.$$

So $ab \geq 3$, $a \geq b$ and $a \geq \sqrt{3} > \frac{17}{10}$.

$a_1, \ldots, a_k, a_{k+1}, \ldots, a_n$ are positive numbers ($k < n$). Suppose that the values of $a_{k+1}, \ldots, a_n$ are fixed. How should one choose the values of $a_1, \ldots, a_n$ in order to minimize $\sum_{i,j,i \neq j} \frac{a_i}{a_j}$?

Solution

To minimize the given rational function, choose

$$a_i = \left(\frac{a_{k+1} + \cdots + a_n}{\frac{1}{a_{k+1}} + \cdots + \frac{1}{a_n}} \right)^{1/2} = (A \cdot H)^{1/2}, \quad i = 1, 2, \ldots, k$$

where A is the arithmetic and H the harmonic mean of $a_{k+1}, \ldots, a_n$.

To prove this, we will be forgiven if we change notation: let $x_i = a_i$, $i = 1, 2, \ldots, k$ and $b_r = a_{k+r}$, $r = 1, \ldots, m$ with $k + m = n$, and denote the given rational function $F(x_1, \ldots, x_k)$. Then we have $F(x_1, \ldots, x_k) = X + Y + B$, where

$$X = \sum_{1 \leq i < j \leq k} \left(\frac{x_i}{x_j} + \frac{x_j}{x_i} \right),$$

$$Y = \sum_{1 \leq i \leq k} \sum_{1 \leq r \leq m} \left(\frac{x_i}{b_r} + \frac{b_r}{x_i} \right),$$

$$B = \sum_{1 \leq r < s \leq m} \left(\frac{b_r}{b_s} + \frac{b_s}{b_r} \right).$$

Note that B is fixed and Y can be improved to

$$Y = \sum_{1 \leq i \leq k} \left(\left(\sum_{1 \leq r \leq m} \frac{1}{b_r} \right) x_i + \left(\sum_{1 \leq r \leq m} b_i \right) \frac{1}{x_i} \right)$$

$$= \sum_i \left(\frac{m}{H} x_i + \frac{mA}{x_i} \right)$$

where A is the arithmetic mean and H is the harmonic mean of the b_r.

Now we recall that the simple function $\alpha x + \frac{\beta}{x}$ (with α, β, x all positive) assumes its minimum when $\alpha x = \frac{\beta}{x}$; that is $x = \sqrt{\beta/\alpha}$. Thus each of the terms in Y (and so Y itself) assumes its minimum when we choose, for $i = 1, 2, \ldots, k$,

$$x_i = \sqrt{\frac{mA}{(m/H)}} = \sqrt{AH}.$$

as asserted.

But there is more. It is also known that each term in X, (and so X itself) assumes its minimum when $x_i = x_j$, with $1 \leq i < j \leq k$. Thus choosing all $x_i = \sqrt{AH}$ minimizes *both* X and Y and, since B is fixed, minimizes $F(x_1, \ldots, x_k)$ as claimed.

m, n are 2 different natural numbers. Show that there exists a real number x, such that $\frac{1}{3} \leq \{xn\} \leq \frac{2}{3}$ and $\frac{1}{3} \leq \{xm\} \leq \frac{2}{3}$, where $\{a\}$ is the fractional part of a.

Solution

We work in the first quadrant of the standard uv-plane, studying the ray
$$R = \{(u, v) = (xm, xn) : x > 0\}.$$

The key is to observe that the problem is equivalent to showing that the ray R contacts at least one of the "small" $\frac{1}{3}$ by $\frac{1}{3}$ squares in the centres of the large standard 1 by 1 lattice squares (considered as a 3×3 checkerboard). (For if (xm, xn) lies in one of these small squares then $(\{xm\}, \{xn\})$ lies in the small square $\{(u, v) : \frac{1}{3} \leq u, v \leq \frac{2}{3}\}$ closest to the origin, as desired.)

To establish this contact, we assume, without loss of generality, that $0 < n < m$, so that R is given by $v = \frac{n}{m}u$, $u > 0$.

Consider the sequence of rays $v = \frac{1}{2}u$, $v = \frac{1}{5}u$, $v = \frac{1}{8}u$, $v = \frac{1}{11}u$, ..., with $u > 0$. These rays are determined by the lower right corners of the 1st, 2nd, 3rd, 4th....., small central square.

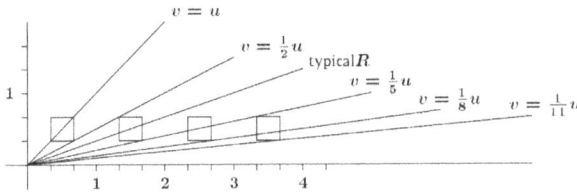

It is now apparent that our ray R lies between, (or on) the ray $v = u$ and $v = \frac{1}{2}u$, or $v = \frac{1}{2}u$ and $v = \frac{1}{5}u$, or ..., and hence R will contact the first or the second or ..., small square, as required.

Determine the number of real solutions a to the equation

$$\left[\frac{1}{2}a\right] + \left[\frac{1}{3}a\right] + \left[\frac{1}{5}a\right] = a.$$

Here, if x is a real number, then $[x]$ denotes the greatest integer that is less than or equal to x.

Solution

Let $a = 30k + r$, where k is an integer and r is a real number between 0 and 29 inclusive.

Then $\left[\frac{1}{2}a\right] = \left[\frac{1}{2}(30k + r)\right] = 15k + \left[\frac{r}{2}\right]$. Similarly $\left[\frac{1}{3}a\right] = 10k + \left[\frac{r}{3}\right]$ and $\left[\frac{1}{5}a\right] = 6k + \left[\frac{r}{5}\right]$.

Now, $\left[\frac{1}{2}a\right] + \left[\frac{1}{3}a\right] + \left[\frac{1}{5}a\right] = a$, so $\left(15k + \left[\frac{r}{2}\right]\right) + \left(10k + \left[\frac{r}{3}\right]\right) + \left(6k + \left[\frac{r}{5}\right]\right) = 30k + r$, and hence $k = r - \left[\frac{r}{2}\right] - \left[\frac{r}{3}\right] - \left[\frac{r}{5}\right]$.

Clearly, r has to be an integer, or $r - \left[\frac{r}{2}\right] - \left[\frac{r}{3}\right] - \left[\frac{r}{5}\right]$ will not be an integer, and therefore, cannot equal k.

On the other hand, if r is an integer, then $r - \left[\frac{r}{2}\right] - \left[\frac{r}{3}\right] - \left[\frac{r}{5}\right]$ will also be an integer, giving exactly one solution for k.

For each r $(0 \leq r \leq 29)$, $a = 30k + r$ will have a different remainder $\mod 30$, so no two different values of r give the same result for a.

Since there are 30 possible values for r $(0, 1, 2, \ldots, 29)$, there are then 30 solutions for a.

Find all real numbers x such that

$$x = \left(x - \frac{1}{x}\right)^{1/2} + \left(1 - \frac{1}{x}\right)^{1/2}.$$

Solution

Since $\left(x - \frac{1}{x}\right)^{1/2} \geq 0$ and $\left(1 - \frac{1}{x}\right)^{1/2} \geq 0$, then $0 \leq \left(x - \frac{1}{x}\right)^{1/2} + \left(1 - \frac{1}{x}\right)^{1/2} = x$.

Note that $x \neq 0$. Else, $\frac{1}{x}$ would not be defined, so $x > 0$.

Squaring both sides gives,

$$x^2 = \left(x - \frac{1}{x}\right) + \left(1 - \frac{1}{x}\right) + 2\sqrt{\left(x - \frac{1}{x}\right)\left(1 - \frac{1}{x}\right)}$$

$$x^2 = x + 1 - \frac{2}{x} + 2\sqrt{x - 1 - \frac{1}{x} + \frac{1}{x^2}}.$$

Multiplying both sides by x and rearranging, we get

$$x^3 - x^2 - x + 2 = 2\sqrt{x^3 - x^2 - x + 1}$$

$$(x^3 - x^2 - x + 1) - 2\sqrt{x^3 - x^2 - x + 1} + 1 = 0$$

$$\left(\sqrt{x^3 - x^2 - x + 1} - 1\right)^2 = 0$$

$$\sqrt{x^3 - x^2 - x + 1} = 1$$

$$x^3 - x^2 - x + 1 = 1$$

$$x(x^2 - x - 1) = 0$$

$$x^2 - x - 1 = 0 \qquad \text{since } x \neq 0.$$

Thus $x = \frac{1 \pm \sqrt{5}}{2}$. We must check to see if these are indeed solutions.

Let $\alpha = \frac{1 + \sqrt{5}}{2}$, $\beta = \frac{1 - \sqrt{5}}{2}$. Note that $\alpha + \beta = 1$, $\alpha\beta = -1$ and $\alpha > 0 > \beta$.

Since $\beta < 0$, β is not a solution.

Now, if $x = \alpha$, then

$$\left(\alpha - \frac{1}{\alpha}\right)^{1/2} + \left(1 - \frac{1}{\alpha}\right)^{1/2}$$

$$= (\alpha + \beta)^{1/2} + (1 + \beta)^{1/2} \quad \text{(since } \alpha\beta = -1\text{)}$$

$$= 1^{1/2} + (\beta^2)^{1/2} \quad \text{(since } \alpha + \beta = 1 \text{ and } \beta^2 = \beta + 1\text{)}$$

$$= 1 - \beta \quad \text{(since } \beta < 0\text{)}$$

$$= \alpha \quad \text{(since } \alpha + \beta = 1\text{)}.$$

So $x = \alpha$ is the unique solution to the equation.

Let n be a natural number such that $n \geq 2$. Show that

$$\frac{1}{n+1}\left(1 + \frac{1}{3} + \cdots + \frac{1}{2n-1}\right) > \frac{1}{n}\left(\frac{1}{2} + \frac{1}{4} + \cdots + \frac{1}{2n}\right).$$

Solution

$$1 + \frac{1}{3} + \ldots + \frac{1}{2n-1} = \frac{1}{2} + \frac{1}{2} + \frac{1}{3} + \frac{1}{5} + \ldots + \frac{1}{2n-1} \qquad (1)$$

Since

$$\frac{1}{3} > \frac{1}{4}, \; \frac{1}{5} > \frac{1}{6}, \; \ldots, \frac{1}{2n-1} > \frac{1}{2n}.$$

(1) gives

$$1 + \frac{1}{3} + \ldots + \frac{1}{2n-1} > \frac{1}{2} + \frac{1}{2} + \frac{1}{4} + \frac{1}{6} \qquad (2)$$
$$+ \ldots + \frac{1}{2n} = \frac{1}{2} + \left(\frac{1}{2} + \frac{1}{4} + \frac{1}{6} + \ldots + \frac{1}{2n}\right).$$

Since

$$\frac{1}{2} > \frac{1}{4}, \; \frac{1}{2} > \frac{1}{6}, \; \frac{1}{2} > \frac{1}{8}, \; \ldots . \; \frac{1}{2} > \frac{1}{2n}$$

then

$$\frac{n}{2} = \underbrace{\frac{1}{2} + \frac{1}{2} + \frac{1}{2} + \ldots + \frac{1}{2}}_{n} > \frac{1}{2} + \frac{1}{4} + \frac{1}{6} + \ldots + \frac{1}{2n}$$

so that

$$\frac{1}{2} > \frac{1}{n}\left(\frac{1}{2} + \frac{1}{4} + \frac{1}{6} + \ldots + \frac{1}{2n}\right). \qquad (3)$$

Then (1), (2) and (3) show that

$$1 + \frac{1}{3} + \ldots + \frac{1}{2n-1}$$
$$> \frac{1}{n}\left(\frac{1}{2} + \frac{1}{4} + \frac{1}{6} + \ldots + \frac{1}{2n}\right) + \left(\frac{1}{2} + \frac{1}{4} + \frac{1}{6} + \ldots + \frac{1}{2n}\right)$$
$$= \left(1 + \frac{1}{n}\right)\left(\frac{1}{2} + \frac{1}{4} + \ldots + \frac{1}{2n}\right)$$
$$= \frac{n+1}{n}\left(\frac{1}{2} + \frac{1}{4} + \ldots + \frac{1}{2n}\right).$$

Therefore $\frac{1}{n+1}\left(1 + \frac{1}{3} + \ldots + \frac{1}{2n-1}\right) > \frac{1}{n}\left(\frac{1}{2} + \frac{1}{4} + \ldots + \frac{1}{2n}\right)$ for all $n \in N$ and $n \geq 2$.

Assume that a and b are integers. Prove that the equation $a^2 + b^2 + x^2 = y^2$ has an integer solution x, y if and only if the product ab is even.

Solution

First, we prove that this condition is necessary. Suppose ab is odd. Then a, b are odd and $a^2 \equiv b^2 \equiv 1 \bmod 4$. Now $x^2 \equiv 0$ or $1 \bmod 4$, and $y^2 \equiv 0$ or $1 \bmod 4$. Therefore $a^2 + b^2 + x^2 = y^2$ is not possible, since if we consider this modulo 4, $2 + x^2 \equiv y^2 \bmod 4$, which is impossible since $2 + x^2 \equiv 2$ or $3 \bmod 4$. Therefore ab must be even.

If ab is even, then, without loss of generality, $a = 2k$.

Consider $4k^2 + b^2 + x^2 = y^2$.

If $4k^2 + b^2 = 2t + 1$, t an integer, then set $y - x = 1$ and $y + x = 2t + 1$, $2y = (t + 1)2$, $y = t + 1$ and $x = t$.

Then $2t + 1 + t^2 = (t + 1)^2$. We are done.

If $4k^2 + b^2$ is even, then $b = 2s$ and $4k^2 + b^2 = 4(k^2 + s^2) = 4m$. Again, $y^2 - x^2 = 4m$.

Set $y - x = 2$ and $y + x = 2m$. Then $y = m + 1$ and $x = y - 2 = m - 1$.

Now $4m + (m - 1)^2 = (m + 1)^2$, and again we are done. Hence $a^2 + b^2 + x^2 = y^2$ always has a solution when ab is even.

To each pair of real numbers a and b, where $a \neq 0$ and $b \neq 0$, there is a real number $a * b$ such that

$$a * (b * c) = (a * b) \cdot c,$$

$$a * a = 1.$$

Solve the equation $x * 36 = 216$.

Solution

Now $a * (a * a) = (a * a) \cdot a$, so

$$a * 1 = 1 \cdot a = a.$$

Also $a * (b * b) = (a * b) \cdot b$ and $a = a * 1 = (a * b) \cdot b$ so

$$a * b = \frac{a}{b}.$$

Finally

$$\frac{x}{36} = 216 \implies x = 7776.$$

Suppose that $yz + zx + xy = 1$ and x, y, and $z \geq 0$. Prove that

$$x(1 - y^2)(1 - z^2) + y(1 - z^2)(1 - x^2) + z(1 - x^2)(1 - y^2) \leq \frac{4\sqrt{3}}{9}.$$

Solution

We first convert the inequality to the following equivalent homogeneous one:

$$x(T_2 - y^2)(T_2 - z^2) + y(T_2 - z^2)(T_2 - x^2) + z(T_2 - x^2)(T_2 - y^2)$$
$$\leq (4\sqrt{3}/9)(T_2)^{5/2}$$

where $T_2 = yz + zx + xy$, and for subsequent use $T_1 = x + y + z$, $T_3 = xyz$. Expanding out, we get

$$T_1 T_2^2 - T_2 \sum x(y^2 + z^2) + T_2 T_3 \leq (4\sqrt{3}/9)(T_2)^{5/2},$$

or

$$T_1 T_2^2 - T_2(T_1 T_2 - 3T_3) + T_2 T_3 = 4T_2 T_3 \leq (4\sqrt{3}/9)(T_2)^{5/2}.$$

Squaring, we get one of the known Maclaurin inequalities for symmetric functions:

$$\sqrt[3]{T_3} \leq \sqrt[2]{T_2/3}.$$

There is equality if and only if $x = y = z$.

Determine all triples of positive rational numbers (x, y, z) such that $x + y + z$, $x^{-1} + y^{-1} + z^{-1}$ and xyz are integers.

Solution

Let $x + y + z = n_1$, $\frac{1}{x} + \frac{1}{y} + \frac{1}{z} = n_2$, and $xyz = n_3$, where n_1, n_2, n_3 are integers. Then $yz + zx + xy = n_2 n_3$ and x, y, z are roots of the cubic

$$t^3 - n_1 t^2 + n_2 n_3 t - n_3 = 0.$$

As known, the only rational roots of the latter are factors of n_3, and consequently x, y, z are integers.

The only triples of integers (x, y, z), aside from permutations, which satisfy $\frac{1}{x} + \frac{1}{y} + \frac{1}{z} = n_2$ are

$$(1, 1, 1), \ (1, 2, 2), \ (2, 3, 6), \ (2, 4, 4), \quad \text{and} \quad (3, 3, 3).$$

Let $A_1, A_2, \ldots, A_8$ be the vertices of a parallelepiped and let O be its centre. Show that

$$4(OA_1^2 + OA_2^2 + \cdots + OA_8^2) \leq (OA_1 + OA_2 + \cdots + OA_8)^2.$$

Solution

Let one of the vertices be the origin and let the vectors $B + C$, $C + A$, $A + B$ denote the three coterminal edges emanating from this origin. Then the vectors to the remaining four vertices are $S + A$, $S + B$, $S + C$, and $2S$ where $S = A + B + C$ and which is also the vector to the centre. The inequality now becomes

$$2(S^2 + A^2 + B^2 + C^2) \leq (|S| + |A| + |B| + |C|)^2.$$

or

$$S^2 + A^2 + B^2 + C^2 \leq 2|S|\{|A| + |B| + |C|\} + 2\{|B|\,|C| + |C|\,|A| + |A|\,|B|\}.$$

Since

$$S^2 = A^2 + B^2 + C^2 + 2B \cdot C + 2C \cdot A + 2A \cdot B.$$

the inequality now becomes

$$S^2 - B \cdot C - C \cdot A - A \cdot B \leq |S|\{|A| + |B| + |C|\} + \{|B|\,|C| + |C|\,|A| + |A|\,|B|\}.$$

Clearly,

$$S^2 \leq |S|\{|A| + |B| + |C|\}$$

and

$$-B \cdot C - C \cdot A - A \cdot B \leq |B|\,|C|\,|A| + |A|\,|B|.$$

There is equality if and only if the parellelepiped is degenerate, for example, $B = C = O$.

Let x, y be positive integers with $y > 3$ and

$$x^2 + y^4 = 2[(x-6)^2 + (y+1)^2].$$

Prove that $x^2 + y^4 = 1994$.

Solution

Rewriting we get

$$x^2 - 24x - y^4 + 2y^2 + 4y + 74 = 0. \qquad (1)$$

Now (1) has integer solutions only if the discriminant $4(y^4 - 2y^2 - 4y + 70)$ is a perfect square. It is easy to prove that for $y \geq 4$,

$$(y^2 - 2)^2 < y^4 - 2y^2 - 4y + 70 < (y^2 + 1)^2. \qquad (*)$$

(Indeed $(*) \iff y^2 - 2y + 33 > 0$ and $4y(y+1) > 69$. The first inequality is true. Since $y \geq 4$, $4y(y+1) \geq 4 \cdot 4 \cdot 5 = 80 > 69$.) The only perfect squares between $(y^2 - 2)^2$ and $(y^2 + 1)^2$ are $(y^2 - 1)^2$ and $(y^2)^2$. Now

$$(y^2 - 1)^2 = y^4 - 2y^2 - 4y + 70 \iff y = \frac{69}{4} \notin \mathbb{Z},$$

and $y^4 - 2y^2 - 4y + 70 = y^4 \iff y^2 + 2y - 35 = 0 \iff y = 5$ or $y = -7$. Thus, $y = 5$. Now (1) gives $x = 37$ and

$$x^2 + y^4 = 37^2 + 5^4 = 1369 + 625 = 1994.$$

A sequence $\{x_n\}$ is defined by the rules

$$x_1 = 2$$

and

$$nx_n = 2(2n-1)x_{n-1}; \qquad n = 2, 3, \ldots.$$

Prove that x_n is an integer for every positive integer n.

Solution

Now

$$x_2 = 2 \cdot \tfrac{3}{2} x_1$$

$$x_3 = 2 \cdot \tfrac{5}{3} x_2$$

$$\cdots$$

$$x_{n-1} = 2 \cdot \tfrac{2n-3}{n-1} x_{n-2}$$

$$x_n = 2 \cdot \tfrac{2n-1}{n} x_{n-1}$$

It follows that

$$x_n = 2^{n-1} \frac{(2n-1)(2n-3)\cdots 5 \cdot 3}{n \cdot (n-1) \cdots 3 \cdot 2} \cdot 2$$

$$= \frac{(2n)!}{(n!)^2} = \binom{2n}{n} \in \mathbb{Z}$$

since $\binom{2n}{n}$ is a binomial coefficient.

Let p, q, r be distinct real numbers which satisfy the equations

$$q = p(4 - p),$$
$$r = q(4 - q),$$
$$p = r(4 - r).$$

Find all possible values of $p + q + r$.

Solution

From the discriminant of each quadratic equation, it follows that p, q, r are all less than 4. If one of p, q, r is zero, all are zero, so we now assume none are zero. Also, it follows that p, q, r all have the same sign. From the product of the three equations we get $1 = (4 - p)(4 - q)(4 - r)$ so that p, q, r are all positive.

We now let $p = 4\sin^2 \theta$, so then successively, $q = 4\sin^2 2\theta$, $r = 4\sin^2 4\theta$, $p = 4\sin^2 8\theta$. Hence, $\sin \theta = \pm \sin 8\theta$ so that we have

$$(\sin 7\theta/2)(\cos 9\theta/2) = 0 \quad \text{or} \quad (\sin 9\theta/2)(\cos 7\theta/2) = 0.$$

Solving for θ leads to only the following possible non-zero values of $p+q+r$: $4(\sin^2 \pi/7 + \sin^2 2\pi/7 + \sin^2 3\pi/7)$, $4(\sin^2 \pi/9 + \sin^2 2\pi/9 + \sin^2 4\pi/9)$ and $3 \cdot 4\sin^2 \pi/3$; that is, 9.

Let w, a, b, c be distinct real numbers with the property that there exist real numbers x, y, z for which the following equations hold:

$$
\begin{aligned}
x + y + z &= 1 \\
xa^2 + yb^2 + zc^2 &= w^2 \\
xa^3 + yb^3 + zc^3 &= w^3 \\
xa^4 + yb^4 + zc^4 &= w^4.
\end{aligned}
$$

Express w in terms of a, b, c.

Solution

As known, the following determinant must vanish:

$$
\begin{vmatrix}
1 & 1 & 1 & 1 \\
a^2 & b^2 & c^2 & w^2 \\
a^3 & b^3 & c^3 & w^3 \\
a^4 & b^4 & c^4 & w^4
\end{vmatrix}
$$

Also, from a known expansion theorem [1] for "alternant" determinants, we have

$$(w - a)(w - b)(w - c)(a - b)(a - c)(b - c)(w(bc + ca + ab) + abc) = 0.$$

Hence, $w = a$, or b, or c, or $-abc/[bc + ca + ab]$.

Let a, b and c be positive real numbers such that $abc = 1$. Prove that

$$\frac{ab}{a^5 + b^5 + ab} + \frac{bc}{b^5 + c^5 + bc} + \frac{ca}{c^5 + a^5 + ca} \leq 1.$$

When does equality hold?

Solution

We note first that

$$\frac{a^5 + b^5}{2} \geq \left(\frac{a^3 + b^3}{2}\right)\left(\frac{a^2 + b^2}{2}\right),$$

since $a^5 - a^3b^2 - a^2b^3 + b^5 = (a - b)^2(a + b)(a^2 + ab + b^2) \geq 0$ with equality if and only if $a = b$.

Similarly

$$\frac{a^3 + b^3}{2} \geq \left(\frac{a + b}{2}\right)\left(\frac{a^2 + b^2}{2}\right)$$

because $a^3 - a^2b - ab^2 + b^3 = (a - b)^2(a + b) \geq 0$, with equality if and only if $a = b$. Thus

$$\frac{a^5 + b^5}{2} \geq \left(\frac{a^3 + b^3}{2}\right)\left(\frac{a^2 + b^2}{2}\right) \geq ab\left(\frac{a^3 + b^3}{2}\right)$$

$$\geq ab\left(\frac{a + b}{2}\right)\left(\frac{a^2 + b^2}{2}\right) \geq \frac{a^2b^2(a + b)}{2}.$$

It is enough, therefore, to prove

$$\frac{ab}{ab(a + b)ab + ab} + \frac{bc}{bc(b + c)bc + bc} + \frac{ca}{ca(c + a)ca + ca} \leq 1,$$

or

$$\frac{1}{ab(a + b) + abc} + \frac{1}{bc(b + c) + abc} + \frac{1}{ca(c + a) + abc} \leq 1.$$

Equivalently,

$$\frac{1}{ab(a + b + c)} + \frac{1}{bc(a + b + c)} + \frac{1}{ca(a + b + c)} \leq 1,$$

or

$$\frac{c}{abc(a + b + c)} + \frac{a}{abc(a + b + c)} + \frac{b}{abc(a + b + c)} \leq 1.$$

Again, because $abc = 1$ we get

$$\frac{a + b + c}{a + b + c} \leq 1,$$

which is true.

The equality requires $a = b = c = 1$.

Find all ordered triples (a, b, c) of real numbers such that for every three integers x, y, z the following identity holds:

$$|ax + by + cz| + |bx + cy + az| + |cx + ay + bz| = |x| + |y| + |z|.$$

Solution

Set $x = y = z = 1$; we obtain $|a + b + c| = 1$ (1)

Set $x = 1$; $y = z = 0$ we obtain $|a| + |b| + |c| = 1$ (2)

Set $x = 1$; $y = -1$, $z = 0$ we obtain $|a - b| + |b - c| + |c - a| = 2$ (3)

This system is symmetric. Without loss of generality we may assume $a \geq b \geq c$.

Now (3) becomes $2(a - c) = 2$ or $a - c = 1$. Substituting into (1) and (2) gives

$$|1 + b + 2c| = 1 \qquad (4)$$

and

$$|1 + c| + |b| + |c| = 1. \qquad (5)$$

Squaring (4) and expanding gives

$$1 + (b + 2c)^2 + 2(b + 2c) = 1.$$

Thus $b + 2c = 0$ or $b + 2c = -2$.

If $b + 2c = 0$, then from (5)

$$|1 + c| + 3|c| = 1.$$

Since $|c| \leq 1$, $1 + c \geq 0$, therefore $1 + c + 3|c| = 1$ and $c + 3|c| = 0$. If $c \geq 0$, we have $4c = 0$ and then $c = 0$. If $c \leq 0$, $-2c = 0$ giving $c = 0$. Therefore $b = -2c = 0$, $a = 1 + c = 1$, in this case.

In case $b + 2c = -2$, substitution into (5) yields

$$|1 + c| + 2|1 + c| + |c| = 1.$$

Since $1 + c \geq 0$, $3(1 + c) + |c| = 1$. If $c \geq 0$, $3 + 4c = 1$ and $c = \frac{-1}{2}$. This is impossible.

If $c \leq 0$, $3 + 3c - c = 1$ giving $c = -1$. Then $b = 0$ and $a = 1 + c = 0$. Therefore we have the solution $a = 0$, $b = 0$, $c = -1$, and these are the solutions for $a \geq b \geq c$.

Hence there are six solutions

$$(1, 0, 0), (-1, 0, 0), (0, 1, 0), (0, -1, 0), (0, 0, 1), (0, 0, -1).$$

Solve the equation

$$\frac{1}{2}(x+y)(y+z)(z+x) + (x+y+z)^3 = 1 - xyz$$

in integers.

Solution

Let $s = x + y + z$ and

$$\begin{aligned} P(X) &= (X-x)(X-y)(X-z) \\ &= X^3 - sX^2 + (xy+yz+zx)X - xyz. \end{aligned}$$

Then $(x+y)(y+z)(z+x) = P(s) = s(xy+yz+xz) - xyz$ and the given equation may be written

$$s(xy+yz+xz) - xyz + 2s^3 = 2 - 2xyz,$$

or $2 + P(-s) = 0$.

As $P(-s) = -(2x+y+z)(2y+z+x)(2z+x+y)$, the equation finally becomes

$$(2x+y+z)(2y+z+x)(2z+x+y) = 2.$$

Either one of the three factors of the left-hand side is 2 and the other two are 1, 1 (or -1, -1) or one of the factors is -2 and the other two are 1, -1, (or -1, 1).

The system

$$\begin{cases} 2x+y+z = 2 \\ x+2y+z = 1 \\ x+y+2z = 1 \end{cases} \qquad \text{is equivalent to} \qquad x=1,\ y=0,\ z=0.$$

The system

$$\begin{cases} 2x+y+z = 2 \\ x+2y+z = -1 \\ x+y+2z = -1 \end{cases} \qquad \text{is equivalent to} \qquad x=2,\ y=-1,\ z=-1.$$

When one of the factors is -2, the two corresponding systems lead to $4(x+y+z) = -2$, which is impossible for integral x, y, z.

Since x, y, z have symmetrical roles, there are six solutions altogether for the triple (x, y, z):

$$(1,0,0),\ (0,1,0),\ (0,0,1),\ (2,-1,-1),\ (-1,2,-1),\ (-1,-1,2).$$

Let $n > 1$ be an odd positive integer. Assume that the integers x_1, $x_2, \ldots, x_n \geq 0$ satisfy the system of equations

$$(x_2 - x_1)^2 + 2(x_2 + x_1) + 1 = n^2,$$
$$(x_3 - x_2)^2 + 2(x_3 + x_2) + 1 = n^2,$$
$$\ldots\ldots\ldots\ldots\ldots\ldots\ldots\ldots\ldots\ldots\ldots$$
$$(x_1 - x_n)^2 + 2(x_1 + x_n) + 1 = n^2.$$

Show that either $x_1 = x_n$, or there exists j with $1 \leq j \leq n - 1$, such that $x_j = x_{j+1}$.

Solution

We show, by induction on m, that, if non-negative integers $x_1, x_2, \ldots,$ $x_m \geq 0$, with $m > 1$ being an odd integer, satisfy

$$(x_2 - x_1)^2 + 2(x_2 + x_1) + 1 = n^2,$$
$$(x_3 - x_2)^2 + 2(x_3 + x_1) + 1 = n^2,$$
$$\ldots\ldots\ldots\ldots\ldots\ldots\ldots\ldots\ldots\ldots\ldots$$
$$(x_1 - x_m)^2 + 2(x_1 + x_m) + 1 = n^2,$$

then either $x_1 = x_m$, or there is j with $1 \leq j \leq m - 1$, such that $x_j = x_{j+1}$. We adopt the convention that subscripts are read modulo m, so that $x_{m+1} = x_1$, etc..

Now notice that for each j, x_{j-1} and x_{j+1} are solutions (possibly equal) to the quadratic equation

$$X^2 + X(2 - 2x_j) + (x_j + 1)^2 - n^2 = 0 \quad (E_j),$$

for which the discriminant $\Delta_j = 4(n^2 - 4x_j) \geq 0$ for there to be a real root. Moreover as n is odd, $\Delta_j \geq 1$.

Also because (E_j) has integral coefficients and at least one integral root, both roots are integers, and they are distinct as $\Delta_j \geq 1$. Denote the roots by α_j and β_j with $\alpha_j < \beta_j$.

Also we have $\alpha_j + \beta_j = 2x_j - 2$, and that $\{x_{j-1}, x_{j+1}\} \subset \{\alpha_j, \beta_j\}$. We claim that *either $\alpha_j = x_j - 2$ and $\beta_j = x_j$ or $\alpha_j < x_j < \beta_j$.* (Indeed if $\alpha_j, \beta_j \geq x_j$ then $\alpha_j + \beta_j > 2x_j - 2$, and if $\alpha_j, \beta_j < x_j$ then $\alpha_j + \beta_j \leq 2x_j - 3$.)

Now let j be such that $x_j = \max\{x_1, \ldots, x_n\}$. Now x_{j-1} and x_{j+1} are roots of E_j. Unless $x_{j-1} = x_j$ or $x_j = x_{j+1}$ we must have

$$x_{j-1} = x_{j+1}$$

or
$$x_{j-1} < x_j < x_{j+1}$$

or
$$x_{j+1} < x_j < x_{j-1}.$$

The last two cases contradict the choice of j, so $x_{j-1} = x_{j+1}$. If $m = 3$ we are done since x_{j-1} and x_{j+1} are cyclically adjacent. Otherwise $m > 3$ and by removing x_{j-1}, x_j we obtain a solution with $m - 2$ values, to which the induction hypothesis applies.

Let n and r be natural numbers. Find the smallest natural number m satisfying this condition: For each partition of the set $\{1, 2, \ldots, m\}$ into r subsets $A_1, A_2, \ldots, A_r$, there exist two numbers a and b in some A_i $(1 \leq i \leq r)$ such that $1 < \frac{a}{b} \leq 1 + \frac{1}{n}$.

Solution

I claim the answer is $(nr + r)$. Suppose $m < nr + r$. Then for each $i \leq m$, let i be put into the subset A_j, $(j = 1, 2, \ldots, r)$ such that $i \equiv j \pmod{r}$. Then for any 2 numbers a, b in the same subset with $a > b$, we have $b \leq (a - r)$. Thus

$$a/b = (1 + (a - b)/b) \geq (1 + r/b) > (1 + r/nr) = (1 + 1/n).$$

Therefore the condition in the question cannot hold. For $m = (nr + r)$, consider the $(r + 1)$ numbers $nr, (nr + 1), \ldots, (nr + r)$. By the Pigeonhole Principle, among them there exists $a > b$ such that a and b are in the same subset. Then

$$a/b = (1 + (a - b)/b) \leq (1 + r/b) \leq (1 + r/nr) = (1 + 1/n),$$

and $1 < a/b$ and we are done.

Solve the following system of equations:

$$x \cdot |x| + y \cdot |y| = 1, \qquad \lfloor x \rfloor + \lfloor y \rfloor = 1,$$

in which $|t|$ and $\lfloor t \rfloor$ represent the absolute value and the integer part of the real number t.

Solution

We show that there are only two solutions: $(x, y) = (1, 0)$ and $(0, 1)$.

Clearly, x and y cannot both be negative. Hence by symmetry, there are two cases to be considered:

(i) If $x \geq 0$ and $y \geq 0$, then from $x^2 + y^2 = 1$ we get $0 \leq x, y \leq 1$. If $x < 1$ and $y < 1$, then $\lfloor x \rfloor + \lfloor y \rfloor = 0$, a contradiction. Hence $x = 1$ or $y = 1$. Then from $x^2 + y^2 = 1$ we obtain the two solutions $(1, 0)$ and $(0, 1)$.

(ii) If $x \geq 0$ and $y < 0$ then $x^2 - y^2 = 1$, and from $\lfloor x \rfloor \leq x$, $\lfloor y \rfloor \leq y$ we get $x + y \geq 1$. Since $x - y > x + y \geq 1$ we have $x^2 - y^2 = (x - y)(x + y) > 1$, a contradiction. Therefore, there are no solutions in this case.

Let $a \in \mathbb{R}$ be given. Find the real numbers $x_1, \ldots, x_n$ which satisfy the system of equations

$$
\begin{aligned}
x_1^2 + ax_1 + \left(\tfrac{a-1}{2}\right)^2 &= x_2, \\
x_2^2 + ax_2 + \left(\tfrac{a-1}{2}\right)^2 &= x_3, \\
&\cdots\cdots\cdots\cdots \\
x_{n-1}^2 + ax_{n-1} + \left(\tfrac{a-1}{2}\right)^2 &= x_n, \\
x_n^2 + ax_n + \left(\tfrac{a-1}{2}\right)^2 &= x_1.
\end{aligned}
$$

Solution

The first equation becomes

$$
x_1^2 + (a-1)x_1 + \left(\frac{a-1}{2}\right)^2 = x_2 - x_1 \quad \text{or} \quad \left(x_1 + \frac{a-1}{2}\right)^2 = x_2 - x_1.
$$

Similarly

$$
\begin{aligned}
\left(x_1 + \tfrac{a-1}{2}\right)^2 &= x_2 - x_1, \\
\left(x_2 + \tfrac{a-1}{2}\right)^2 &= x_3 - x_2, \\
&\cdots\cdots\cdots\cdots \\
\left(x_{n-1} + \tfrac{a-1}{2}\right)^2 &= x_n - x_{n-1}, \\
\left(x_n + \tfrac{a-1}{2}\right)^2 &= x_1 - x_n.
\end{aligned}
$$

Adding, we get:

$$
\left(x_1 + \tfrac{a-1}{2}\right)^2 + \left(x_2 + \tfrac{a-1}{2}\right)^2 + \cdots + \left(x_{n-1} + \tfrac{a-1}{2}\right)^2 + \left(x_n + \tfrac{a-1}{2}\right)^2 = 0.
$$

So $x_1 = x_2 = \cdots = x_{n-1} = x_n = \frac{1-a}{2}$.

Show that $\cos(\sin x) > \sin(\cos x)$ holds for every real number x.

Solution

First Solution. $-\frac{\pi}{2} < -1 \leq \sin x \leq 1 < \frac{\pi}{2} \implies \cos(\sin x) > 0$ for $x \in \mathbb{R}$.

If $\cos x \leq 0$, then $\cos x \in (-\frac{\pi}{2}, 0]$, so $\sin(\cos x) \leq 0 < \cos(\sin x)$.

If $\cos x > 0$, then $\cos x \in (0, \frac{\pi}{2}]$. It is known that $\sin y < y$ for $y \in (0, \frac{\pi}{2})$. So

$$\sin(\cos x) < \cos x. \tag{1}$$

Also $\cos y \geq 1 - \frac{y^2}{2}$ for $y \in (-\frac{\pi}{2}, \frac{\pi}{2})$. Since $\sin x \in (-\frac{\pi}{2}, \frac{\pi}{2})$ we have

$$\cos(\sin x) \geq 1 - \frac{\sin^2 x}{2} = \frac{1 + \cos^2 x}{2}. \tag{2}$$

But, using (1) and (2), $\dfrac{1 + \cos^2 x}{2} \geq \cos x \implies \cos(\sin x) > \sin(\cos x)$.

Second Solution.

$$\cos(\sin x) - \sin(\cos x)$$
$$= \cos(\sin x) - \cos\left(\frac{\pi}{2} - \cos x\right)$$
$$= 2\sin\left(\frac{\sin x - \cos x + \frac{\pi}{2}}{2}\right)\sin\left(\frac{\frac{\pi}{2} - \sin x - \cos x}{2}\right)$$
$$= 2\sin\left(\frac{\sqrt{2}}{2}\sin\left(x - \frac{\pi}{4}\right) + \frac{\pi}{4}\right)\left(\sin\left(\frac{\pi}{4} - \frac{\sqrt{2}}{2}\sin\left(x + \frac{\pi}{4}\right)\right)\right).$$

It is easy to prove that

$$0 < \frac{\pi}{4} - \frac{\sqrt{2}}{2} \leq \frac{\sqrt{2}}{2}\sin\left(x - \frac{\pi}{4}\right) + \frac{\pi}{4},$$

$$\frac{\pi}{4} - \frac{\sqrt{2}}{2}\sin\left(x + \frac{\pi}{4}\right) \leq \frac{\pi}{4} + \frac{\sqrt{2}}{2} < \frac{\pi}{2},$$

so that

$$\sin\left(\frac{\sqrt{2}}{2}\sin\left(x - \frac{\pi}{4}\right) + \frac{\pi}{4}\right), \quad \sin\left(\frac{\pi}{4} - \frac{\sqrt{2}}{2}\sin\left(x + \frac{\pi}{4}\right)\right) > 0.$$

Consider all pairs of real numbers satisfying the inequalities

$$-1 \leq x + y \leq 1, \quad -1 \leq xy + x + y \leq 1.$$

Let M denote the largest possible value of x.

(a) Prove that $M \leq 3$.

(b) Prove that $M \leq 2$.

(c) Find M.

Solution

Letting $u = x + 1$ and $v = y + 1$, the given inequalities become

$$0 \leq uv \leq 2, \quad 1 \leq u + v \leq 3.$$

Clearly u and v must each be non-negative and so the maximum value of u is 3. Finally, $x_{\max} = 2$.

Find all real solutions to the equation $4x^2 - 40[x] + 51 = 0$. Here, if x is a real number, then $[x]$ denotes the greatest integer that is less than or equal to x.

Solution

Rearranging the equation we get $4x^2 + 51 = 40[x]$. It is known that $x \geq [x] > x - 1$, so

$$4x^2 + 51 = 40[x] > 40(x - 1),$$
$$4x^2 - 40x + 91 > 0,$$
$$(2x - 13)(2x - 7) > 0.$$

Hence $x > 13/2$ or $x < 7/2$. Also,

$$4x^2 + 51 = 40[x] \leq 40x,$$
$$4x^2 - 40x + 51 \leq 0,$$
$$(2x - 17)(2x - 3) \leq 0.$$

Hence $3/2 \leq x \leq 17/2$. Combining these inequalities gives $3/2 \leq x < 7/2$ or $13/2 < x \leq 17/2$.

Case 1: $3/2 \leq x < 7/2$.

For this case, the possible values for $[x]$ are 1, 2 and 3.

If $[x] = 1$ then $4x^2 + 51 = 40 \cdot 1$ so $4x^2 = -11$, which has no real solutions.

If $[x] = 2$ then $4x^2 + 51 = 40 \cdot 2$ so $4x^2 = 29$ and $x = \frac{\sqrt{29}}{2}$. Notice that $\frac{\sqrt{16}}{2} < \frac{\sqrt{29}}{2} < \frac{\sqrt{36}}{2}$ so $2 < x < 3$ and $[x] = 2$.

If $[x] = 3$ then $4x^2 + 51 = 40 \cdot 3$ and $x = \sqrt{69}/2$. But $\frac{\sqrt{69}}{2} > \frac{\sqrt{64}}{2} = 4$. So, this solution is rejected.

Case 2: $13/2 < x \leq 17/2$.

For this case, the possible values for $[x]$ are 6, 7 and 8.

If $[x] = 6$ then $4x^2 + 51 = 40 \cdot 6$, so that $x = \frac{\sqrt{189}}{2}$. Notice that $\frac{\sqrt{144}}{2} < \frac{\sqrt{189}}{2} < \frac{\sqrt{196}}{2}$, so that $6 < x < 7$ and $[x] = 6$.

If $[x] = 7$ then $4x^2 + 51 = 40 \cdot 7$, so that $x = \frac{\sqrt{229}}{2}$. Notice that $\frac{\sqrt{196}}{2} < \frac{\sqrt{229}}{2} < \frac{\sqrt{256}}{2}$, so that $7 < x < 8$ and $[x] = 7$.

If $[x] = 8$ then $4x^2 + 51 = 40 \cdot 8$, so that $x = \frac{\sqrt{269}}{2}$. Notice that $\frac{\sqrt{256}}{2} < \frac{\sqrt{269}}{2} < \frac{\sqrt{324}}{2}$, so that $8 < x < 9$ and $[x] = 8$.

The solutions are $x = \dfrac{\sqrt{29}}{2}, \dfrac{\sqrt{189}}{2}, \dfrac{\sqrt{229}}{2}, \dfrac{\sqrt{269}}{2}$.

Suppose a_1, a_2, ..., a_8 are eight distinct integers from $\{1, 2,$ ..., $16, 17\}$. Show that there is an integer $k > 0$ such that the equation $a_i - a_j = k$ has at least three different solutions. Also, find a specific set of 7 distinct integers from $\{1, 2, \ldots, 16, 17\}$ such that the equation $a_i - a_j = k$ does not have three distinct solutions for any $k > 0$.

Solution

Without loss of generality let $a_1 < a_2 < a_3 \ldots < a_8$.

Assume that there is no such integer k. Let us just look at the seven differences $d_i = a_{i+1} - a_i$. Then amongst the d_i there can be at most two 1s, two 2s, and two 3s, which totals to 12.

Now $16 = 17 - 1 \geq a_8 - a_1 = d_1 + d_2 + \ldots + d_7$ so the seven differences must be 1, 1, 2, 2, 3, 3, 4.

Now let us think of arranging the differences 1, 1, 2, 2, 3, 3, 4. Note that the sum of consecutive differences is another difference. (For example, $d_1 + d_2 = a_3 - a_1$, $d_1 + d_2 + d_3 = a_4 - a_1$)

We cannot place the two 1s side by side because that will give us another difference of 2. The 1s cannot be beside a 2 because then we have three 3s. They cannot both be beside a 3 because then we have three 4s! So we must have either $1, 4, -, -, -, 3, 1$ or $1, 4, 1, 3, -, -, -$ (or their reflections).

In either case we have a 3, 1 giving a difference of 4 so we cannot put the 2s beside each other. Also we cannot have 2, 3, 2 because then (with the 1, 4) we will have three 5s. So all cases give a contradiction.

Therefore there will always be three differences equal.

One set of seven numbers satisfying the criteria is $\{1, 2, 4, 7, 11, 16, 17\}$. [*Editor*: There are many such sets.]

Solution 2

Consider all the consecutive differences (that is, d_i above) as well as the differences $b_i = a_{i+2} - a_i$, $i = 1, \ldots, 6$. Then the sum of these thirteen differences is $2 \cdot (a_8 - a_1) + (a_7 - a_2) \leq 2(17 - 1) + (16 - 2) = 46$. Now if no difference occurs more than twice, the smallest the sum of the thirteen differences can be is $2 \cdot (1 + 2 + 3 + 4 + 5 + 6) + 7 = 49$, giving a contradiction.

Let x, y, and z be non-negative real numbers satisfying $x + y + z = 1$. Show that

$$x^2 y + y^2 z + z^2 x \leq \frac{4}{27},$$

and find when equality occurs.

Solution

Let $f(x, y, z) = x^2 y + y^2 z + z^2 x$. We wish to determine where f is maximal. Since f is cyclic, without loss of generality we may assume that $x \geq y, z$. Since

$$\begin{aligned} f(x, y, z) - f(x, z, y) &= x^2 y + y^2 z + z^2 x - x^2 z - z^2 y - y^2 x \\ &= (y - z)(x - y)(x - z), \end{aligned}$$

we may also assume $y \geq z$. Then

$$\begin{aligned} f(x + z, y, 0) - f(x, y, z) &= (x + z)^2 y - x^2 y - y^2 z - z^2 x \\ &= z^2 y + yz(x - y) + xz(y - z) \geq 0, \end{aligned}$$

so we may now assume $z = 0$. The rest follows from the arithmetic-geometric mean inequality:

$$f(x, y, 0) = \frac{2x^2 y}{2} \leq \frac{1}{2}\left(\frac{x + x + 2y}{3}\right)^3 = \frac{4}{27}.$$

Equality occurs when $x = 2y$, hence at $(x, y, z) = (\frac{2}{3}, \frac{1}{3}, 0)$. (As well as $(0, \frac{2}{3}, \frac{1}{3})$ and $(\frac{1}{3}, 0, \frac{2}{3})$).

Solution 2
With f as above, and $x \geq y, z$

$$f\left(x + \frac{z}{2}, y + \frac{z}{2}, 0\right) - f(x, y, z) = yz(x - y) + \frac{xz}{2}(x - z) + \frac{z^2 y}{4} + \frac{z^3}{8},$$

so we may assume that $z = 0$. The rest follows as for solution 1.

For a given positive integer m, find all pairs (n, x, y) of positive integers such that m, n are relatively prime and $(x^2 + y^2)^m = (xy)^n$, where n, x, y can be represented by functions of m.

Solution

Let p be a common prime divisor of x and y, and let α and β be the largest integers such that $p^\alpha \mid x$ and $p^\beta \mid y$.

Now $p^{2\alpha} \mid x^2$, $p^{2\beta} \mid y^2$ and

$$p^{(\alpha+\beta)n} \mid (xy)^n = (x^2 + y^2)^m.$$

We claim that $\alpha = \beta$ (and $x = y$). Indeed, if $\alpha < \beta$, then $p^{2\alpha} \mid x^2 + y^2$ and $p^{2\alpha m} \mid (x^2 + y^2)^m$; that is

$$2\alpha m = (\alpha + \beta)n > 2\alpha n,$$

and then $m > n$. But this is impossible since

$$(xy)^m < (2xy)^m \leq (x^2 + y^2)^m = (xy)^n.$$

Similarly if $\alpha > \beta$ we obtain a contradiction. Hence $x = y$ and $(x^2+y^2)^m = (xy)^n$ reduces to $2^m x^{2m} = x^{2n}$. The solutions are of the form $(n, x, y) = (m + 1, 2^{m/2}, 2^{m/2})$.

If m is odd there is no solution. For m even $n = m + 1$, $x = y = 2^m$.

Moreover, if we do not suppose $(n, m) = 1$ the solutions are those numbers of the form $m = 2a\alpha$, $n = m + \alpha$, $x = y = 2^a$.

For any positive integer m, show that there exist integers a, b satisfying

$$|a| \leq m, \quad |b| \leq m, \quad 0 < a + b\sqrt{2} \leq \frac{1 + \sqrt{2}}{m + 2}.$$

Solution

When a and b run through $0, 1, \ldots, m$, then $a + b\sqrt{2}$ takes $(m+1)^2$ distinct values and the largest one is $m + m\sqrt{2}$. Dividing $[0, m + m\sqrt{2}]$ into $m^2 + 2m$ intervals of length $\frac{1+\sqrt{2}}{m+2}$ and using the Pigeonhole Principle, we deduce the existence of two elements (a_1, b_1), (a_2, b_2) with $a_1 + b_1\sqrt{2} > a_2 + b_2\sqrt{2} > 0$ and $\left(a_1 + b_1\sqrt{2}\right) - \left(a_2 + b_2\sqrt{2}\right) < \frac{1+\sqrt{2}}{m+2}$. Hence $a := a_1 - a_2$, $b := b_1 - b_2$ satisfy the required condition.

Solve the system

$$x + \log(x + \sqrt{x^2 + 1}) = y\,,$$
$$y + \log(y + \sqrt{y^2 + 1}) = z\,,$$
$$z + \log(z + \sqrt{z^2 + 1}) = x\,.$$

Solution

It is clear that $x = y = z = 0$ is a solution and if any one of x, y, and z is zero, then they are all zeros. Assume then $xyz \neq 0$. Note that if t is a real number such that $t > 0$, then $t + \sqrt{t^2 + 1} > 1 \implies \log(t + \sqrt{t^2 + 1}) > 0$. On the other hand, if $t < 0$, then

$$-2t > 0 \implies t^2 + 1 < (1 - t)^2$$
$$\implies \sqrt{t^2 + 1} < 1 - t \implies t + \sqrt{t^2 + 1} < 1$$
$$\implies \log(t + \sqrt{t^2 + 1}) < 0\,.$$

Label the three given equations as (1), (2), and (3). If $x > 0$, then we get $y > x > 0$ by (1), $z > y > 0$ by (2) and $x > z$ by (3). Thus $x > z > y > x$, which is a contradiction. Similarly, if $x < 0$, then we get $y < x < 0$ by (1), $z < y < 0$ by (2) and $x < z$ by (3). Thus $x < z < y < x$, which is again a contradiction.

Therefore, $x = y = z = 0$ is the only solution.

Prove for each choice of real non-zero numbers a_1, a_2, a_3, the "stars" can be replaced by "$<$" and "$>$" so that the system

$$\begin{cases} a_1 x + b_1 y + c_1 * 0 & (1) \\ a_2 x + b_2 y + c_2 * 0 & (2) \\ a_3 x + b_3 y + c_3 * 0 & (3) \end{cases}$$

has no solution.

Solution

Let Δ_i be the line with equation $a_i x + b_i y + c_i = 0$ (we have $(a_i, b_i) \neq (0, 0)$).

Let π_i^+ be the open half-plane with equation $a_i x + b_i y + c_i > 0$ and π_i^- be the open half-plane with equation $a_i x + b_i y + c_i < 0$.

Case 1. Two of the lines are parallel.

Without loss of generality, we may take Δ_1 and Δ_2 to be parallel.

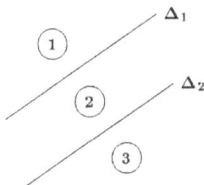

We can select the $*$ for (1) so that (1) is the region and for (2) so that (3) is determined.

Then there is no point in the intersection and the system has no solution (even if $\Delta_1 = \Delta_2$ because the inequalities are strict).

Case 2. No two of the lines are parallel.

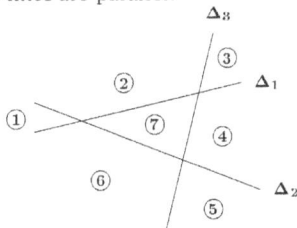

The intersections between $\pi_i^{\mp}$ and $\pi_j^{\pm}$ for $i \neq j$ determine 7 regions. We replace the stars in (1) and (3) to determine the region labelled (3). Then replace the star in (2) so that the solution must lie on the opposite side of Δ_2 (that is, (1), (6) or (5)). Then the system has no solution.

Solve in natural numbers:

$$x(x+1) = y^7.$$

Solution

We show more generally that if $n > 1$ is a natural number, then $x(x+1) = y^n$ has no solutions in natural numbers. Clearly, $y \neq 1$. Let p denote any prime divisor of y. Then we must have $p^n \mid x$ or $p^n \mid x+1$ as x and $x+1$ are coprime. Thus x and $x+1$ must each be an nth power. That is, $x = b^n$, $x+1 = a^n$ for some natural numbers a and b with $a > b$. If $n > 1$, then from $a^n - b^n = 1$ we get $(a-b)(a^{n-1}+a^{n-2}b+\cdots+b^{n-1}) = 1$, which is clearly impossible since $a - b \geq 1$ and $a^{n-1} + a^{n-2}b + \cdots + b^{n-1} > 1$.

Solve the equation $\cos x \cdot \cos 2x \cdot \cos 3x = 1$.

Solution

Letting $y = \cos 2x$ and using the formula

$$2\cos A \cos B = \cos(A+B) + \cos(A-B) \quad \text{we have}$$

$$\cos x \cdot \cos 2x \cdot \cos 3x = 1$$

$$\Longleftrightarrow (\cos 2x)(\cos 4x + \cos 2x) = 2$$

$$\Longleftrightarrow (\cos 2x)(2\cos^2 2x - 1 + \cos 2x) = 2$$

$$\Longleftrightarrow 2y^3 + y^2 - y - 2 = 0$$

$$\Longleftrightarrow (y-1)(2y^2 + 3y + 2) = 0$$

$$\Longleftrightarrow y = 1 \quad (\because 2y^2 + 3y + 2 \text{ has no real roots})$$

$$\Longleftrightarrow \cos 2x = 1 \Longleftrightarrow 2x = 2k\pi \Longleftrightarrow x = k\pi,$$

where k is any integer.

Find all pairs of positive integers x, y such that

$$x^2 + 615 = 2^y.$$

Solution

The only solution is $x = 59$ and $y = 12$. Note first that for all non-negative integers k

$$2^{2k+1} \equiv 4^k \cdot 2 \equiv (-1)^k \cdot 2 \equiv 2 \text{ or } 3 \pmod{5}.$$

Since $x^2 \equiv 0$, 1 or $4 \pmod 5$, y must be even. Letting $y = 2z$, then $x^2 + 615 = 2^y$ becomes $(2^z - x)(2^z + x) = 615$. Since $615 = 3 \times 5 \times 41$ there are only 4 cases:

(1) $2^z + x = 615$, $2^z - x = 1$;

(2) $2^z + x = 205$, $2^z - x = 3$;

(3) $2^z + x = 123$, $2^z - x = 5$;

(4) $2^z + x = 41$, $2^z - x = 15$.

Cases (1), (2) and (4) clearly yield no solution, since $2^{z+1} = 616$, 208 or 56, none of which is a power of two. Finally, case (3) yields $2^{z+1} = 128$ or $z = 6$, from which it follows that $x = 59$ and $y = 12$.

Prove that for any natural number n, the average of all its factors lies between the numbers $\sqrt{n}$ and $\frac{n+1}{2}$.

Solution

Let $d_1, d_2, \ldots, d_k$ denote the positive divisors of n where $k = \tau(n)$ is the number of (positive) divisors of n. We are to show that

$$\sqrt{n} \leq \frac{1}{k}\sum_{i=1}^{k} d_i \leq \frac{n+1}{2}. \tag{1}$$

To establish the right inequality in (1) we first show that

$$d + \frac{n}{d} \leq n+1 \tag{2}$$

for all divisors d of n.

This is clearly true if $d = 1$. If $d > 1$, then

$$d + \frac{n}{d} \leq n+1 \iff d^2 + n \leq d(n+1)$$
$$\iff d(d-1) \leq (d-1)n \iff d \leq n,$$

which clearly holds. Since $d \mid n \iff \frac{n}{d} \mid n$ we have, from (2) that

$$2\sum_{i=1}^{k} d_i = \sum_{i=1}^{k}\left(d_i + \frac{n}{d_i}\right) \leq k(n+1)$$

from which $\frac{1}{k}\sum_{i=1}^{k} d_i \leq \frac{n+1}{2}$ follows.

If equality holds, then for *any* divisor d of n, either $d = 1$ or $\frac{n}{d} = 1$. Thus $k = 2$ and n must be a prime. Conversely, if n is a prime, then $k = 2$ and $\frac{1}{k}\sum_{i=1}^{k} d_i = \frac{n+1}{2}$. To show the left inequality in (1) we use the Arithmetic-Geometric-Mean Inequality. Note that if n is not a perfect square, then for all divisors d of n we have $d \neq \frac{n}{d}$ and so k must be even. Pairing off d with $\frac{n}{d}$, we obtain $\prod_{i=1}^{k} d_i = n^{k/2}$. If $n = q^2$ is a perfect square, then k is odd. Again, pairing off d with $\frac{n}{d}$ for all $d \neq q$, we find that

$$\prod_{i=1}^{k} d_i = q \cdot n^{(k-1)/2} = n^{1/2} \cdot n^{(k-1)/2} = n^{k/2},$$

which is the same as above.

Hence in both cases, we have

$$\frac{1}{k}\sum_{i=1}^{k} d_i \geq \left(\prod_{i=1}^{k} d_i\right)^{1/k} = (n^{k/2})^{1/k} = \sqrt{n}.$$

Clearly, equality holds in this inequality if and only if $n = 1$.

Denote the sum of the first n prime numbers by S_n. Prove that there exists a whole square between S_n and S_{n+1}.

Solution

We have $S_1 < 2^2 < S_2 < 3^2 < S_3 < 4^2 < S_4 < 5^2 < S_5$.

Suppose that $n \geq 5$, so that $p_n \geq 11$. Let a_n be the integer ≥ 6 defined by $p_n = 2a_n - 1$, and S'_n the sum of all odd integers from 1 to p_n, inclusive. It is well known that $S'_n = a_n^2 = \left(\frac{p_n+1}{2}\right)^2$ and easily seen that $S'_n > S_n$, because $n \geq 5$. Now let us assume that there is no square between S_n and S_{n+1}. Then there would exist an integer k such that $k^2 \leq S_n < S_{n+1} \leq (k+1)^2$ and we would have

$$S_{n+1} - S_n \leq 2k + 1; \quad \text{that is } p_{n+1} \leq 2k + 1 .$$

From this we could write successively:

$$p_n \leq 2k - 1, \quad k \geq \frac{p_n + 1}{2} \quad \text{and} \quad S_n \geq \left(\frac{p_n + 1}{2}\right)^2$$

which contradicts $S'_n > S_n$. The result follows.

Show that, if $(x + \sqrt{x^2 + 1})(y + \sqrt{y^2 + 1}) = 1$, then $x + y = 0$.

Solution 1

We prove, more generally, that if $(x + \sqrt{x^2 + 1})(y + \sqrt{y^2 + 1}) = p$, then $x + y = \frac{p-1}{\sqrt{p}}$.

Set $z = x + \sqrt{x^2 + 1}$. Then $z > 0$ and $x = \frac{z^2 - 1}{2z}$. Consequently, we obtain

$$y + \sqrt{y^2 + 1} = \frac{p}{z}$$

and then

$$y = \frac{(p/z)^2 - 1}{(2p/z)} = \frac{p^2 - z^2}{2pz}.$$

Hence,

$$x + y = \frac{z^2 - 1}{2z} + \frac{p^2 - z^2}{2zp} = \frac{p-1}{2p}\left(z + \frac{p}{z}\right)$$

$$\geq \frac{p-1}{p}\sqrt{z \cdot \frac{p}{z}} = \frac{p-1}{\sqrt{p}}.$$

Equality occurs for $x = y = \frac{p-1}{2\sqrt{p}}$.

Solution 2

Taking logarithms, the hypothesis implies

$$\ln(x + \sqrt{x^2 + 1}) + \ln(y + \sqrt{y^2 + 1}) = 0 \quad \text{or}$$

$$\sinh^{-1}(x) + \sinh^{-1}(y) = 0.$$

Since the function $\sinh^{-1}$ is odd we obtain

$$\sinh^{-1}(x) = \sinh^{-1}(-y) \quad \text{and} \quad x = -y,$$

as required.

Find the smallest natural number m such that, for all natural numbers $n \geq m$, we have $n = 5a + 11b$, with a, b integers ≥ 0.

Solution

We prove the more general result that if $p, q \in \mathbb{N}$ are relatively prime, then the smallest integer m such that, for all $N \geq m$, n can be written as a non-negative integer linear combination of p and q is $m = pq - p - q + 1$. For the present problem $\{p, q\} = \{5, 11\}$ and so the answer is $m = 40$.

For convenience, call an integer n "*expressible*" if $n = ap + bq$ for some non-negative integers a and b. We prove the following result which was first obtained by J.J. Sylvester in 1884 in response to a problem proposed earlier by G. Frobenius.

Theorem. The smallest integer m such that n is expressible for all $n \geq m$ is
$$m = pq - p - q + 1 .$$

Proof. We first show that $m - 1$ is not expressible. Suppose, to the contrary, that $pq - p - q = ap + bq$ where a, b are non-negative integers. Then we have $pq = (1+a)p + (1+b)q$. Hence $p \mid 1+b$ and $q \mid 1+a$. Letting $1 + a = a'q$ and $1 + b = b'p$ where $a' \geq 1$, $b' \geq 1$ we get $pq = (a' + b')pq$ and so $a' + b' = 1$, which is clearly a contradiction.

Next we show that m is expressible. Since $(p, q) = 1$, there exist integers x and y such that $xp + yq = 1$ and thus $(x - kq)p + (y + kp)q = 1$ for all integers k. Since clearly $p \nmid y$, we can choose an appropriate k so that $-p < y + kp < 0$. Then clearly $x - kq > 0$. Let $x_0 = x - kq$ and $y_0 = y + kp$. Then we have $-p < y_0 < 0 < x_0$, $x_0 p + y_0 q = 1$ and so
$$m = pq - p - q + (x_0 p + y_0 q) = (x_0 - 1)p + (p + y_0 - 1)q ,$$
showing that m is indeed expressible.

Now we show by induction that n is expressible for all $n \geq m$. Suppose n_0 is expressible for some $n_0 \geq m$. Then $n_0 = \alpha p + \beta q$ for some non-negative integers α and β, and so
$$n_0 + 1 = (\alpha + x_0)p + (\beta + y_0)q .$$

If $\beta + y_0 \geq 0$, then we are done. Suppose $\beta + y_0 < 0$. Then we write $n_0 + 1 = (\alpha + x_0 - q)p + (\beta + y_0 + p)q$. Since $\beta + y_0 + p > 0$, it remains to show that $\alpha + x_0 - q \geq 0$. If $\alpha + x_0 - q < 0$, then $\alpha + x_0 - q \leq -1$ and thus
$$\alpha + x - kq - q + 1 \leq 0 . \tag{1}$$

On the other hand, since $\beta + y_0 < 0$, we have $\beta + y_0 \leq -1$ and thus
$$\beta + y + kp + 1 < 0 . \tag{2}$$

From $p \times (1) + q \times (2)$ we get
$$\alpha p + \beta q + xp + yq - pq + p + q \leq 0$$
or
$$n_0 + 1 \leq pq - p - q = m - 1 ,$$
a contradiction. Therefore, $\alpha + x_0 - q \geq 0$ and thus $n_0 + 1$ is expressible, completing our proof.

Find all the integer solutions of the equation

$$p(x + y) = xy$$

in which p is a prime number.

Solution

$p(x + y) = xy$ is equivalent to $(x - p)(y - p) = p^2$.

Since p is a prime, the integral divisors of p^2 are $-p^2, -p, -1, 1, p, p^2$. It follows that (x, y) is a solution if and only if $(x, y) \in \{(p(1 - p), p - 1),$ $(0, 0), (p - 1, p(1 - p)), (p + 1, p(p + 1)), (2p, 2p), (p(p + 1), p + 1)\}$.

Show that, if the equations

$$x^3 + mx - n = 0, \quad nx^3 - 2m^2x^2 - 5mnx - 2m^3 - n^2 = 0 \quad (m \neq 0, n \neq 0)$$

have a common root, then the first equation would have two equal roots, and determine in this case the roots of both equations in terms of n.

Solution

Let a be a common root of the two equations. We have $a^3 = n - ma$. Thus, we also have

$$n(n - ma) - 2m^2a^2 - 5mna - 2m^3 - n^2 = 0$$

and a is a root of $\qquad mx^2 + 3nx + m^2 = 0$. $\qquad (1)$

Substituting a for x in

$$x^3 + mx - n = \left(x^2 + 3\frac{n}{m}x + m\right)\left(x - \frac{3n}{m}\right) + \frac{9n^2}{m^2}x + 2n,$$

we obtain $a = -\frac{2m^2}{9n}$, and returning to (1), we deduce that $4m^3 + 27n^2 = 0$. From this, we easily get the following factorization:
$x^3 + mx - n = \left(x - \frac{3n}{2m}\right)^2\left(x + \frac{3n}{m}\right)$, which shows that the first of the given equations has a simple root $-\frac{3n}{m}$ and a double root $\frac{3n}{2m}$.

We note in passing that $a = \frac{3n}{2m}$ (since $4m^3 + 27n^2 = 0$).

The second equation may be rewritten as $nx^3 - 2m^2x^2 - 5mnx + \frac{25}{2}n^2 = 0$, so that the sum of the roots is $\frac{2m^2}{n}$ and the product of the roots is $-\frac{25n}{2}$. If we denote by b and c the roots other than a, we thus have $b + c = \frac{2m^2}{n} - \frac{3n}{2m} = -\frac{15n}{m}$ and $bc = -\frac{25n}{2} \times \frac{2m}{3n} = -\frac{25m}{3}$, from which we easily obtain $b = c = -\frac{15n}{2m}$.

Using $4m^3 + 27n^2 = 0$, we see that $\frac{n}{m} = -\frac{4^{1/3}n^{1/3}}{3}$ (where $x \mapsto x^{1/3}$ denotes the inverse function of $x \mapsto x^3$ from $\mathbb{R}$ to $\mathbb{R}$) and we calculate:

roots of the first equation: $(4n)^{1/3}$; $\quad -\left(\frac{n}{2}\right)^{1/3}$; $\quad -\left(\frac{n}{2}\right)^{1/3}$

roots of the second equation: $-\left(\frac{n}{2}\right)^{1/3}$; $\quad 5\left(\frac{n}{2}\right)^{1/3}$; $\quad 5\left(\frac{n}{2}\right)^{1/3}$.

Let a be a fixed whole number.

Determine all solutions x, y, z in whole numbers to the system of equations

$$\begin{array}{rcrcrcl}
5x & + & (a+2)y & + & (a+2)z & = & a\,, \\
(2a+4)x & + & (a^2+3)y & + & (2a+2)z & = & 3a-1\,, \\
(2a+4)x & + & (2a+2)y & + & (a^2+3)z & = & a+1\,.
\end{array}$$

Solution

The second and third equations give

$$(a-1)^2(y-z) = 2(a-1)\,.$$

When $a = 1$, the original system reduces to two independent equations $5x + 3y + 3z = 1$ and $6x + 4y + 4z = 2$, from which we obtain a one-parameter set of whole-number solutions $x = -1$, $y = 1 + t$, $z = 1 - t$, where $t \in Z$.

When $a \neq 1$, we have $y - z = \dfrac{2}{a-1}$. If y, z are whole numbers, then so is their difference, which means that $a - 1 \mid 2$, restricting the possibilities to $a = -1$, $a = 0$, $a = 2$, $a = 3$.

Case 1. $a = -1$. The equations become

$$\begin{array}{rcl}
5x + y + z & = & -1\,, \\
2x + 4y & = & -4\,, \\
2x + 4z & = & 0\,,
\end{array}$$

giving a solution $x = 0$, $y = -1$, $z = 0$.

Case 2. $a = 0$. The equations become

$$\begin{array}{rcl}
5x + 2y + 2z & = & 0\,, \\
4x + 3y + 2z & = & -1\,, \\
4x + 2y + 3z & = & 1\,,
\end{array}$$

giving a solution $x = 0$, $y = -1$, $z = 1$.

Case 3. $a = 2$. The equations become

$$\begin{array}{rcl}
5x + 4y + 4z & = & 2\,, \\
8x + 7y + 6z & = & 5\,, \\
8x + 6y + 7z & = & 3\,,
\end{array}$$

giving a solution $x = -6$, $y = 5$, $z = 3$.

Case 4. $a = 3$. The equations include

$$5x + 5y + 5z = 3\,,$$

which evidently has no whole-number solutions, since $5 \nmid 3$.

Determine all quadruples (a, b, c, d) of real numbers satisfying the equation

$$256a^3b^3c^3d^3 = (a^6 + b^2 + c^2 + d^2)(a^2 + b^6 + c^2 + d^2)$$
$$\times (a^2 + b^2 + c^6 + d^2)(a^2 + b^2 + c^2 + d^6).$$

Solution

Let (a, b, c, d) be a quadruple of real numbers satisfying the equation. If one of the numbers is zero, then all are 0. From now on, we suppose that none of the four numbers is 0. Since the right-hand side is positive there must be an even number of negative reals amongst a, b, c, d. Then (a, b, c, d) is a solution if and only if $(|a|, |b|, |c|, |d|)$ is a solution. Thus, we may suppose that a, b, c, d are positive.

From the AM–GM Inequality,

$$a^6 + b^2 + c^2 + d^2 \geq 4(a^6b^2c^2d^2)^{1/4},$$

and similarly for the other three factors on the right-hand side of the equation. Thus,

$$(a^6 + b^2 + c^2 + d^2)(a^2 + b^6 + c^2 + d^2)$$
$$\times (a^2 + b^2 + c^6 + d^2)(a^2 + b^2 + c^2 + d^6)$$
$$\geq 256(a^6b^2c^2d^2)^{1/4}(a^2b^6c^2d^2)^{1/4} \times (a^2b^2c^6d^2)^{1/4}(a^2b^2c^2d^6)^{1/4}$$
$$= 256a^3b^3c^3d^3,$$

which indicates that the given equation is the equality case of the AM/GM Inequality. Therefore,

$$a^6 = b^2 = c^2 = d^2 = a^2 = b^6 = c^6 = d^6;$$

that is, $a = b = c = d = 1$.

Then the solutions are $(0, 0, 0, 0)$ and $(\varepsilon_1, \varepsilon_2, \varepsilon_3, \varepsilon_4)$ where $\varepsilon_i = \pm 1$ for $i = 1, 2, 3, 4$ and $\prod_{i=1}^{4} \varepsilon_i = 1$.

Suppose that $w_1, w_2, \ldots, w_k$ are distinct real numbers with a non-zero sum. Prove that there exist integer numbers $n_1, n_2, \ldots, n_k$ such that $\sum_{i=1}^{k} n_i w_i > 0$ and for any non-identity permutation π on $\{1, 2, \ldots, k\}$ we have $\sum_{i=1}^{k} n_i w_{\pi(i)} < 0$.

Solution

First, we prove the following version of the Hardy-Pólya-Littlewood Inequality.

Theorem. Suppose that $a_1 < a_2 < \cdots < a_n$ and $b_1 < b_2 < \cdots < b_n$ are real numbers. Define

$$\alpha = \min_{1 \leq i < n} (a_{i+1} - a_i) \quad \text{and} \quad \beta = \min_{1 \leq i < n} (b_{i+1} - b_i).$$

Then, for any permutation $\pi \neq 1$, we have

$$\sum_{i=1}^{n} b_i a_{\pi(i)} \leq \sum_{i=1}^{n} b_i a_i - \alpha\beta.$$

Proof. Define $A_\pi = \sum_{i=1}^{n} b_i a_{\pi(i)}$. Let σ be a non-identity permutation with maximum value A_σ. There exist $i < j$ such that $\sigma(i) > \sigma(j)$. Set $\sigma' = \sigma \circ (i\ j)$. Then

$$A_{\sigma'} = A_\sigma + (a_{\sigma(i)} - a_{\sigma(j)})(b_j - b_i) \geq A_\sigma + \alpha\beta.$$

Thus, $\sigma' = 1$ and the theorem follows.

Without loss of generality, we can assume $w_1 < w_2 < \cdots < w_k$. Define $\alpha = \min_{1 \leq i < k} (w_{i+1} - w_i)$ and $s = \left| \sum_{i=1}^{k} w_i \right| > 0$. Select a natural number $N > \dfrac{s}{\alpha}$, and set

$$(n_1, n_2, \ldots, n_k) = (N, 2N, \ldots, kN) + p(1, 1, \ldots, 1),$$

where p is the unique integer such that $\sum_{i=1}^{k} n_i w_i \in (0, s]$.

Now, we have

$$N = \min_{1 \leq i < k} (n_{i+1} - n_i) \quad \text{and} \quad \alpha = \min_{1 \leq i < k} (w_{i+1} - w_i).$$

By the above theorem, for $\pi \neq 1$,

$$\sum_{i=1}^{k} n_i w_{\pi(i)} \leq \sum_{i=1}^{k} n_i w_i - N\alpha \leq s - N\alpha < 0.$$

Thus, the proof is complete.

Prove the inequality

$$\frac{1+ab}{1+a} + \frac{1+bc}{1+b} + \frac{1+ac}{1+c} \geq 3$$

for positive real numbers a, b, c with $abc = 1$.

Solution

Since $abc = 1$, we have

$$\frac{1+ab}{1+a} = \frac{1+\frac{1}{c}}{1+a} = \frac{1+c}{c(1+a)}.$$

It follows that

$$\frac{1+ab}{1+a} + \frac{1+bc}{1+b} + \frac{1+ac}{1+c} = \frac{1}{c} \cdot \frac{1+c}{1+a} + \frac{1}{a} \cdot \frac{1+a}{1+b} + \frac{1}{b} \cdot \frac{1+b}{1+c}$$

$$\geq 3\sqrt[3]{\frac{1}{c} \cdot \frac{1+c}{1+a} \cdot \frac{1}{a} \cdot \frac{1+a}{1+b} \cdot \frac{1}{b} \cdot \frac{1+b}{1+c}} = 3,$$

using the AM-GM Inequality, and recalling that $abc = 1$.

For real numbers x, y, $z \in (0, 1]$, prove the inequality

$$\frac{x}{1+y+zx} + \frac{y}{1+z+xy} + \frac{z}{1+x+yz} \leq \frac{3}{x+y+z}.$$

Solution

We will prove the stronger inequality

$$\frac{x}{1+y+zx} + \frac{y}{1+z+xy} + \frac{z}{1+x+yz} \leq 1.$$

Since $1 + xy = (1-x)(1-y) + x + y$, we have

$$\begin{aligned} 1 + z + xy &= (1-x)(1-y) + x + y + z \\ &\geq x + y + z. \end{aligned}$$

Analogously, we also have

$$\begin{aligned} 1 + x + yz &= (1-y)(1-z) + x + y + z \geq x + y + z \\ \text{and} \quad 1 + y + zx &= (1-z)(1-x) + x + y + z \geq x + y + z. \end{aligned}$$

Therefore,

$$\begin{aligned} &\frac{x}{1+y+zx} + \frac{y}{1+z+xy} + \frac{z}{1+x+yz} \\ &\leq \frac{x}{x+y+z} + \frac{y}{x+y+z} + \frac{z}{x+y+z} = 1, \end{aligned}$$

as claimed.

Let $x_1, x_2, \ldots, x_n, \ldots$ be the sequence of real numbers such that

$$x_1 = 1, \quad x_{n+1} = \frac{n^2}{x_n} + \frac{x_n}{n^2} + 2, \quad n \geq 1.$$

Prove that

(a) $x_{n+1} \geq x_n$ for all $n \geq 4$;

(b) $[x_n] = n$ for all $n \geq 4$ ($[a]$ denotes the whole part of a).

Solution

We can easily compute the first few terms:

$$x_1 = 1, \quad x_2 = 4, \quad x_3 = 4, \quad x_4 = \frac{169}{36}.$$

By induction on $n \geq 4$, we shall prove the inequalities

$$n + \frac{2}{n} < x_n < n + 1. \tag{1}$$

The initial case, $n = 4$, is easy to check. For the induction step, we must show that (1) implies

$$n + 1 + \frac{2}{n+1} < x_{n+1} < n + 2. \tag{2}$$

Let

$$f_n(x) = \frac{n^2}{x} + \frac{x}{n^2} + 2.$$

Then

$$x_{n+1} = f_n(x_n), \quad n \geq 1. \tag{3}$$

The function $f_n(x)$ has derivative $f_n'(x) = (x^2 - n^4)/(n^2 x^2)$. Hence, it is decreasing on $(0, n^2)$. Using (1) and (3), we have

$$f_n(n+1) < f_n(x_n) = x_{n+1} < f_n\left(n + \frac{2}{n}\right). \tag{4}$$

Also

$$
\begin{aligned}
f_n(n+1) &= \frac{n^2}{n+1} + \frac{n+1}{n^2} + 2 \\
&= n + 1 + \frac{1}{n+1} + \frac{n+1}{n^2} > n + 1 + \frac{2}{n+1},
\end{aligned} \tag{5}
$$

and

$$f_n\left(n + \frac{2}{n}\right) = f_n\left(\frac{n^2 + 2}{n}\right) = \frac{n^3}{n^2 + 2} + \frac{n^2 + 2}{n^3} + 2.$$

Note that

$$f_n\left(n + \frac{2}{n}\right) < n + 2 \tag{6}$$

for $n \geq 3$, as a consequence of

$$\frac{n^3}{n^2+2} + \frac{n^2+2}{n^3} + 2 - (n+2) = \frac{-\left[(n^2-2)^2 - 8\right]}{n^3(n^2+2)} < 0.$$

From (5), (4), and (6), we obtain

$$n+1+\frac{2}{n+1} < f_n(n+1) < x_{n+1} < f_n\left(n+\frac{2}{n}\right) < n+2,$$

and thus (2) is proved.

Therefore, (1) is true for all $n \geq 4$. Hence, $[x_n] = n$ for all $n \geq 4$, and part (b) is proved.

Now we consider part (a). We have

$$x_{n+1} - x_n = \frac{n^2}{x_n} + \frac{x_n}{n^2} + 2 - x_n = \frac{(n^2+x_n)^2 - (nx_n)^2}{n^2 x_n}.$$

For $n \geq 4$, we have, from (1), $x_n < n+1 < \frac{n^2}{n-1}$, which implies

$$nx_n < x_n + n^2.$$

Hence, $x_{n+1} > x_n$, for $n \geq 4$. [*Ed.* From above we also have $x_{n+1} \geq x_n$ for $n = 1, 2$, and 3, which means that $x_{n+1} \geq x_n$ for all $n \geq 1$.]

Let $x_1, x_2, \ldots, x_n$ $(n \geq 2)$ be real positive numbers satisfying

$$\frac{1}{x_1 + 1998} + \frac{1}{x_2 + 1998} + \cdots + \frac{1}{x_n + 1998} = \frac{1}{1998}.$$

Prove that

$$\frac{\sqrt[n]{x_1 x_2 \cdots x_n}}{n - 1} \geq 1998.$$

Solution

Let $y_j = x_j / 1998$ $(j = 1, 2, \ldots, n)$. Then the hypothesis reads

$$\frac{1}{1 + y_1} + \frac{1}{1 + y_2} + \cdots + \frac{1}{1 + y_n} = 1.$$

Now, this implies $y_1 y_2 \cdots y_n \geq (n - 1)^n$

Let n be an integer, $n \geq 2$, and $0 < a_1 < a_2 < \cdots < a_{2n+1}$ be real numbers. Prove that the following inequality holds:

$$\sqrt[n]{a_1} - \sqrt[n]{a_2} + \sqrt[n]{a_3} - \cdots - \sqrt[n]{a_{2n}} + \sqrt[n]{a_{2n+1}}$$
$$< \sqrt[n]{a_1 - a_2 + a_3 - \cdots - a_{2n} + a_{2n+1}} \,.$$

Solution

Let $n \geq 2$ be an integer.

Lemma. If $0 \leq a < b \leq c < d$ are real numbers such that $a^n + d^n = b^n + c^n$, then $a + d < b + c$.

Proof. Let $x = b - a$, $y = c - b$, and $z = d - c$. From the Binomial Theorem, we have

$$\begin{aligned}
a^n + (a + x + y + z)^n &= a^n + d^n = b^n + c^n \\
&= (a + x)^n + (a + x + y)^n \\
&= \sum_{k=0}^{n} \binom{n}{k} a^{n-k} (x^k + (x+y)^k) \\
&< 2a^n + \sum_{k=1}^{n} \binom{n}{k} a^{n-k} (2x + y)^k \\
&= a^n + (a + 2x + y)^n \,.
\end{aligned}$$

Then $z < x$; that is $a + d < b + c$. The lemma is proved.

Let $0 < a_1 < a_2 < \cdots < a_{2n+1}$ be real numbers. From the lemma, we deduce that

$$\sqrt[n]{a_1} + \sqrt[n]{a_3 - \cdots - a_{2n} + a_{2n+1}}$$
$$< \sqrt[n]{a_2} + \sqrt[n]{a_1 - a_2 + a_3 - \cdots - a_{2n} + a_{2n+1}} \,,$$
$$\sqrt[n]{a_3} + \sqrt[n]{a_5 - \cdots - a_{2n} + a_{2n+1}}$$
$$< \sqrt[n]{a_4} + \sqrt[n]{a_3 - a_4 + \cdots - a_{2n} + a_{2n+1}} \,,$$

$$\vdots$$

$$\sqrt[n]{a_{2n-1}} + \sqrt[n]{a_{2n+1}} < \sqrt[n]{a_{2n}} + \sqrt[n]{a_{2n-1} - a_{2n} + a_{2n+1}} \,.$$

Summing, we get the desired result.

Prove that the equation $y^2 = x^5 - 4$ has no integer solutions.

Solution

We have $(x^5)^2 = x^{10} \equiv 0$ or $1 \bmod 11$ for all x [by Fermat's Theorem].
Hence, $x^5 \equiv -1$, 0, or $1 \bmod 11$. Thus, $x^5 - 4 \equiv 6$, 7, or $8 \bmod 11$.

But all squares are congruent to 0, 1, 3, 4, 5, or $9 \bmod 11$. Therefore, the equation has no solutions in integers.

If x, y, $z > 0$, $k > 2$ and $a = x + ky + kz$, $b = kx + y + kz$, $c = kx + ky + z$, show that

$$\frac{x}{a} + \frac{y}{b} + \frac{z}{c} \geq \frac{3}{2k+1}.$$

Solution

More generally, if $x_0, x_1, \ldots, x_n > 0$ ($n \geq 1$), $k > 1$, and

$$a_i = x_i + k \sum_{\substack{j=0 \\ j \neq i}}^{n} x_j \quad (i = 0, 1, \ldots, n),$$

then

$$\sum_{i=0}^{n} \frac{x_i}{a_i} \geq \frac{n+1}{nk+1}.$$

Proof. Let $S = \sum_{i=0}^{n} x_i$. Then, for each i we have $a_i = kS - (k-1)x_i$; that is,

$$x_i = \frac{kS - a_i}{k-1}.$$

For all i, $j \in \{0, 1, \ldots, n\}$,

$$\frac{x_i}{a_i} \leq \frac{x_j}{a_j} \iff \frac{kS - a_i}{a_i} \leq \frac{kS - a_j}{a_j} \iff a_j \leq a_i.$$

Hence,

$$\sum_{i=0}^{n} \sum_{j=0}^{n} \left(\frac{x_i}{a_i} - \frac{x_j}{a_j} \right) (a_j - a_i) \geq 0,$$

or, equivalently,

$$\left(\sum_{i=0}^{n} \frac{x_i}{a_i} \right) \left(\sum_{i=0}^{n} a_i \right) \geq (n+1)S.$$

The stated inequality now follows by noting that $\sum_{i=0}^{n} a_i = (nk+1)S$.

To solve the present proposal, take $n = 2$ and rename $x_0 = x$, $x_1 = y$, and $x_2 = z$.

Find all solutions in the real domain of the equation

$$x \cdot \lfloor x \cdot \lfloor x \cdot \lfloor x \rfloor \rfloor \rfloor = 88,$$

where $\lfloor a \rfloor$ is the integer part of a real number a; that is, the integer satisfying $\lfloor a \rfloor \le a < \lfloor a \rfloor + 1$. For instance, $\lfloor 3.7 \rfloor = 3$, $\lfloor -3.7 \rfloor = -4$ and $\lfloor 6 \rfloor = 6$.

Solution

A straightforward calculation shows that $x = \frac{22}{7}$ is a solution. We will prove that this is the only solution.

Let x be a solution and let $n = \lfloor x \rfloor$. Then $n \le x < n + 1$.

Case 1. $x > 0$.

Then

$$n^2 \le x \cdot \lfloor x \rfloor < n^2 + n,$$
$$n^2 \le \lfloor x \cdot \lfloor x \rfloor \rfloor < n^2 + n,$$
$$n^3 \le x \cdot \lfloor x \cdot \lfloor x \rfloor \rfloor < n(n+1)^2,$$
$$n^3 \le \lfloor x \cdot \lfloor x \cdot \lfloor x \rfloor \rfloor \rfloor < n(n+1)^2,$$
$$n^4 \le x \cdot \lfloor x \cdot \lfloor x \cdot \lfloor x \rfloor \rfloor \rfloor < n(n+1)^3,$$
$$n^4 \le 88 < n(n+1)^3.$$

The inequalities in the last line are true only if $n = 3$. Now the equation becomes $x \cdot \lfloor x \cdot \lfloor 3x \rfloor \rfloor = 88$.

Let $k = \lfloor 3x \rfloor$. Then $k \le 3x < k + 1$. Thus,

$$\frac{k}{3} \le x < \frac{k+1}{3},$$
$$\frac{k^2}{3} \le x \cdot \lfloor 3x \rfloor < \frac{k(k+1)}{3},$$
$$\frac{k^2}{3} - 1 < \lfloor x \cdot \lfloor 3x \rfloor \rfloor < \frac{k(k+1)}{3},$$
$$\frac{k^3 - 3k}{9} < x \cdot \lfloor x \cdot \lfloor 3x \rfloor \rfloor < \frac{k(k+1)^2}{9},$$
$$\frac{k^3 - 3k}{9} < 88 < \frac{k(k+1)^2}{9}.$$

These last inequalities are true only if $k = 9$. Now the equation becomes $x \lfloor 9x \rfloor = 88$.

Let $\ell = \lfloor 9x \rfloor$. Then $\ell \leq 9x < \ell + 1$. Hence,

$$\frac{\ell}{9} \leq x < \frac{\ell+1}{9}\,,$$

$$\frac{\ell^2}{9} \leq x \cdot \lfloor 9x \rfloor < \frac{\ell^2 + \ell}{9}\,,$$

$$\frac{\ell^2}{9} \leq 88 < \frac{\ell^2 + \ell}{9}\,,$$

$$\ell^2 \leq 792 < \ell^2 + \ell\,.$$

These inequalities are true only if $\ell = 28$. Finally, from the equation $x \lfloor 9x \rfloor = 88$, we get $x = \dfrac{88}{28} = \dfrac{22}{7}$.

Case 2. $x < 0$.

Then (since $n < 0$) we have

$$n^2 \geq x \cdot \lfloor x \rfloor > n^2 + n\,,$$

$$n^2 \geq \lfloor x \cdot \lfloor x \rfloor \rfloor \geq n^2 + n\,,$$

$$n^3 \leq x \cdot \lfloor x \cdot \lfloor x \rfloor \rfloor \leq n(n+1)^2\,,$$

$$n^3 \leq \lfloor x \cdot \lfloor x \cdot \lfloor x \rfloor \rfloor \rfloor \leq n(n+1)^2\,,$$

$$n^4 \geq x \cdot \lfloor x \cdot \lfloor x \cdot \lfloor x \rfloor \rfloor \rfloor \geq n(n+1)^3\,,$$

$$n^4 \geq 88 \geq n(n+1)^3\,.$$

No integer $n < 0$ satisfies these inequalities.

Let a, b, c be positive numbers. Show that the triangle with sides a, b, c exists if and only if the system of equations

$$\frac{y}{z} + \frac{z}{y} = \frac{a}{x}, \quad \frac{z}{x} + \frac{x}{z} = \frac{b}{y}, \quad \frac{x}{y} + \frac{y}{x} = \frac{c}{z}$$

has a solution in the real domain.

Solution

Suppose first that the given system has a solution (x, y, z), where x, y, and z are (non-zero) real numbers. Then,

$$\begin{aligned}
b + c - a &= y\left(\frac{z}{x} + \frac{x}{z}\right) + z\left(\frac{x}{y} + \frac{y}{x}\right) - x\left(\frac{y}{z} + \frac{z}{y}\right) \\
&= \frac{2yz}{x} = \frac{2(y^2 + z^2)}{a} > 0.
\end{aligned}$$

Thus, $b + c > a$. Similarly, $c + a > b$ and $a + b > c$. Therefore, a, b, c are the sides of some triangle.

Conversely, if a, b, c are the sides of a triangle, let $s = \frac{a + b + c}{2}$ and

$$x = \sqrt{(s - b)(s - c)}, \quad y = \sqrt{(s - c)(s - a)}, \quad z = \sqrt{(s - a)(s - b)}.$$

Then (x, y, z) is a solution of the system. Indeed,

$$\frac{y}{z} + \frac{z}{y} = \frac{\sqrt{s - c}}{\sqrt{s - b}} + \frac{\sqrt{s - b}}{\sqrt{s - c}} = \frac{s - c + s - b}{x} = \frac{a}{x},$$

and the other two equations are similarly verified.

Solve the equation

$$\frac{1}{x^2} + \frac{1}{(4 - \sqrt{3}\,x)^2} = 1\,.$$

Solution

For any solution x, there is some number θ such that $\cos\theta = 1/x$ and $\sin\theta = \dfrac{1}{4 - \sqrt{3}x}$; hence, there is some θ such that

$$4 = \sqrt{3}\sec\theta + \csc\theta\,. \tag{1}$$

Conversely, if θ satisfies the above equation, then, letting $x = \sec\theta$, we have $4 - \sqrt{3}x = \csc\theta$, and x is a solution of the given equation.

Equation (1) is equivalent to

$$2\sin\theta\cos\theta = \tfrac{\sqrt{3}}{2}\sin\theta + \tfrac{1}{2}\cos\theta\,;$$

that is,

$$\sin(2\theta) = \sin\left(\theta + \frac{\pi}{6}\right)\,.$$

This equation is satisfied either when $2\theta = \theta + \frac{\pi}{6} + 2k\pi$ or when $2\theta = (2k+1)\pi - (\theta + \frac{\pi}{6})$, where $k \in \mathbb{Z}$. Thus, the solutions of (1) are given by

$$\theta = \frac{\pi}{6} + 2k\pi \tag{2}$$

or

$$\theta = \frac{5\pi}{18} + \frac{2k\pi}{3}\,. \tag{3}$$

Due to the periodicity of the secant function, all values of θ given by (2) yield $x = \sec\left(\frac{\pi}{6}\right) = \frac{2}{\sqrt{3}} = \frac{2\sqrt{3}}{3}$. Three more distinct values of x are obtained from the values of θ given by (3). These are $x = \sec\left(\frac{5\pi}{18}\right)$, $x = \sec\left(\frac{17\pi}{18}\right)$, and $x = \sec\left(\frac{29\pi}{18}\right)$.

Find the smallest integer n for which $0 < \sqrt[4]{n} - \lfloor \sqrt[4]{n} \rfloor < 10^{-5}$.

Solution

Let $f(n) = n^{1/4} - \lfloor n^{1/4} \rfloor$, for positive integers n. If $n = x^4$ for some positive integer x, then $f(n) = x - x = 0$. Moreover, $f(n)$ is clearly increasing on $[x^4, (x+1)^4)$. Hence, in order to get the *smallest* integer n such that $0 < f(n) < 10^{-5}$, we must choose $n = x^4 + 1$, with x as small as possible.

Now,

$$f(x^4 + 1) = (1 + x^4)^{1/4} - x = x\left(1 + \frac{1}{x^4}\right)^{1/4} - x,$$

which, by the Binomial Theorem, yields

$$f(x^4 + 1) = x\left(1 + \frac{1}{4}\frac{1}{x^4} + \frac{1}{4}\left(\frac{-3}{4}\right)\frac{1}{2!}\frac{1}{x^8} + \cdots\right) - x$$

$$= \frac{1}{4x^3} - \frac{3}{32}\frac{1}{x^7} + \cdots$$

Therefore, we look for x such that $\frac{1}{4x^3} < 10^{-5}$; that is, $4x^3 > 10^5$. Since $4 \times 30^3 = 1.08 \times 10^5$, and a value of $x = 30$ makes the subsequent terms in the binomial expansion negligible compared with 10^{-5}, it follows that $x = 30$ and $n = 30^4 + 1 = 810001$.

Suppose that $a_1 < a_2 < \cdots < a_n$ are real numbers. Prove that:

$$a_1 a_2^4 + a_2 a_3^4 + \cdots + a_{n-1} a_n^4 + a_n a_1^4 \geq a_2 a_1^4 + a_3 a_2^4 + \cdots + a_n a_{n-1}^4 + a_1 a_n^4 .$$

Solution

We prove the result by induction on n. For $n = 2$, we have equality. The case $n = 3$ will be needed below. For $n = 3$, we have to show that

$$a_1 a_2^4 + a_2 a_3^4 + a_3 a_1^4 \geq a_2 a_1^4 + a_3 a_2^4 + a_1 a_3^4 .$$

This is true, since

$$
\begin{aligned}
a_1 a_2^4 &+ a_2 a_3^4 + a_3 a_1^4 - a_2 a_1^4 - a_3 a_2^4 - a_1 a_3^4 \\
&= \tfrac{1}{2}(a_2 - a_1)(a_3 - a_2)(a_3 - a_1) \\
&\qquad \cdot \left[(a_1 + a_2)^2 + (a_2 + a_3)^2 + (a_3 + a_1)^2\right] \geq 0 .
\end{aligned}
$$

Assume that the claim is true for $n - 1$, and let us prove it for n. By applying the induction hypothesis, we find that it is sufficient to prove that

$$a_{n-1} a_n^4 + a_n a_1^4 - a_{n-1} a_1^4 \geq a_n a_{n-1}^4 + a_1 a_n^4 - a_1 a_{n-1}^4 ,$$

which is the case $n = 3$.

Suppose that n is a positive integer and $d_1 < d_2 < d_3 < d_4$ are the four smallest positive integers dividing n. Find all integers n satisfying $n = d_1^2 + d_2^2 + d_3^2 + d_4^2$.

Solution

If n is odd, then $d_1^2 + d_2^2 + d_3^2 + d_4^2 \equiv 1 + 1 + 1 + 1 \equiv 0 \pmod 4$, and we cannot have $n = d_1^2 + d_2^2 + d_3^2 + d_4^2$. Thus, we can assume that 2 divides n. Then $d_1 = 1$ and $d_2 = 2$, and hence,

$$n \equiv 1 + 0 + d_3^2 + d_4^2 \not\equiv 0 \pmod 4 .$$

Thus, $4 \nmid n$.

Hence, $(d_1, d_2, d_3, d_4) = (1, 2, p, q)$ or $(d_1, d_2, d_3, d_4) = (1, 2, p, 2p)$ for some odd primes p, q. In the first case, $n \equiv 3 \pmod 4$, a contradiction. Thus, $n = 5(1 + p^2)$ and $5 \mid n$. Therefore, $p = d_3 = 5$ and $n = 130$.

Suppose that $r_1, \ldots, r_n$ are real numbers. Prove that there exists $I \subseteq \{1, 2, \ldots, n\}$ such that I meets $\{i, i+1, i+2\}$ in at least one and at most two elements, for $1 \le i \le n - 2$ and

$$\left| \sum_{i \in I} r_i \right| \ge \frac{1}{6} \sum_{i=1}^{n} |r_i| .$$

Solution

Let $r = \sum_{i=1}^{n} |r_i|$. For $i = 0, 1, 2$, define

$$s_i = \sum_{\substack{r_j \ge 0 \\ j \equiv i \pmod 3}} r_j \quad \text{and} \quad t_i = \sum_{\substack{r_j < 0 \\ j \equiv i \pmod 3}} r_j .$$

Then we have $r = s_0 + s_1 + s_2 - t_0 - t_1 - t_2$, and

$$\begin{aligned} 2r &= (s_0 + s_1) + (s_1 + s_2) + (s_2 + s_0) \\ &\quad - (t_0 + t_1) - (t_1 + t_2) - (t_2 + t_0) . \end{aligned}$$

Therefore, there exist i_1 and i_2 with $i_1 \ne i_2$ such that $s_{i_1} + s_{i_2} \ge \frac{1}{3}r$ or $t_{i_1} + t_{i_2} \le -\frac{1}{3}r$. Assume, without loss of generality, that

$$s_{i_1} + s_{i_2} \ge \frac{1}{3}r \quad \text{and} \quad s_{i_1} + s_{i_2} \ge -(t_{i_1} + t_{i_2}) .$$

Then $s_{i_1} + s_{i_2} + t_{i_1} + t_{i_2} \ge 0$, and we have

$$(s_{i_1} + s_{i_2} + t_{i_1}) + (s_{i_1} + s_{i_2} + t_{i_2}) \ge s_{i_1} + s_{i_2} \ge \frac{1}{3}r .$$

Hence, one of $s_{i_1} + s_{i_2} + t_{i_1}$ and $s_{i_1} + s_{i_2} + t_{i_2}$ must be at least $\frac{1}{6}r$.

Let m be a given integer. Prove that there exist integers a, b, and k such that both a, b are not divisible by 2, $k \geq 0$, and

$$2m \;=\; a^{19} + b^{99} + k \cdot 2^{1999} \,.$$

Solution

Let n be a positive integer, and let r be an odd positive integer. For any odd positive integers x and y

$$x^r \;\equiv\; y^r \;(\mathrm{mod}\ 2^n) \qquad \Longleftrightarrow \qquad x \;\equiv\; y \;(\mathrm{mod}\ 2^n)\,,$$

because

$$x^r - y^r \;=\; (x - y)\,(x^{r-1} + x^{r-2}y + \cdots + y^{r-1})$$

and $x^{r-1} + x^{r-2}y + \cdots + y^{r-1}$ is odd. It follows that the set of congruence classes of 1^r, 3^r, 5^r, ..., $(2^n - 1)^r$ modulo 2^n is the same as the set of congruence classes of $1, 3, 5, \ldots, 2^n - 1$, which is the set of all odd congruence classes modulo 2^n.

Taking $r = 19$ and $n = 1999$, we see that there exists an odd number a_0 such that $2m - 1 \equiv a_0^{19} \;(\mathrm{mod}\ 2^{1999})$. Choose $a \equiv a_0 \;(\mathrm{mod}\ 2^{1999})$ to be sufficiently negative so that $2m - 1 - a^{19} > 0$. Then a solution is

$$(a, b, k) \;=\; \left(a, 1, \frac{2m - 1 - a^{19}}{2^{1999}}\right)\,.$$

Solve the system of equations

$$\begin{cases} (1 + 4^{2x-y})5^{1-2x+y} & = & 1 + 2^{2x-y+1} \\ y^3 + 4x + 1 + \ln(y^2 + 2x) & = & 0 \end{cases}$$

Solution

Letting $t = 2x - y$, the first equation becomes $f(t) = 0$, where

$$f(t) = 1 + 2 \cdot 2^t - 5 \cdot \left(\tfrac{1}{5}\right)^t - 5 \cdot \left(\tfrac{4}{5}\right)^t.$$

Since the function f is strictly increasing, the obvious solution $t = 1$ is the only solution. Thus, $2x - y = 1$; that is, $y = 2x - 1$.

The second equation then becomes $g(x) = 0$, where

$$g(x) = (2x - 1)^3 + 4x + 1 + \ln(4x^2 - 2x + 1).$$

We easily find

$$g'(x) = 6(2x - 1)^2 + \frac{16x^2 + 2}{3x^2 + (x - 1)^2} > 0.$$

Therefore, g is strictly increasing, and the obvious solution $x = 0$ is the only solution. Finally, $(x, y) = (0, -1)$.

Let $\{x_n\}_{n=0}^{\infty}$ and $\{y_n\}_{n=0}^{\infty}$ be two sequences defined recursively as follows:

$$x_0 = 1, \quad x_1 = 4, \quad x_{n+2} = 3x_{n+1} - x_n,$$
$$y_0 = 1, \quad y_1 = 2, \quad y_{n+2} = 3y_{n+1} - y_n,$$

for all $n = 0, 1, 2, \ldots$.

(a) Prove that

$$x_n^2 - 5y_n^2 + 4 = 0$$

for all non-negative integers n.

(b) Suppose that a, b are two positive integers such that $a^2 - 5b^2 + 4 = 0$. Prove that there exists a non-negative integer k such that $x_k = a$ and $y_k = b$.

Solution

(a) The sequences $\{x_n\}$, $\{y_n\}$ may be defined equivalently by

$$x_0 = y_0 = 1, \quad \begin{pmatrix} x_{n+1} \\ y_{n+1} \end{pmatrix} = A \begin{pmatrix} x_n \\ y_n \end{pmatrix}, \qquad (1)$$

where A denotes the matrix $\begin{pmatrix} 3/2 & 5/2 \\ 1/2 & 3/2 \end{pmatrix}$. Indeed, (1) gives $x_1 = 4$, $y_1 = 2$,

$$\begin{aligned} x_{n+2} &= \tfrac{3}{2}x_{n+1} + \tfrac{5}{2}y_{n+1} = \tfrac{3}{2}x_{n+1} + \tfrac{5}{2}\left(\tfrac{1}{2}x_n + \tfrac{3}{2}y_n\right) \\ &= \tfrac{3}{2}x_{n+1} + \tfrac{5}{4}x_n + \tfrac{3}{2}\left(x_{n+1} - \tfrac{3}{2}x_n\right) = 3x_{n+1} - x_n, \end{aligned}$$

and $y_{n+2} = 3y_{n+1} - y_n$ (similarly). Now, the required result follows by induction, using the relations $x_0^2 - 5y_0^2 + 4 = 0$ and

$$\begin{aligned} x_{n+1}^2 - 5y_{n+1}^2 + 4 &= \left(\tfrac{3}{2}x_n + \tfrac{5}{2}y_n\right)^2 - 5\left(\tfrac{1}{2}x_n + \tfrac{3}{2}y_n\right)^2 + 4 \\ &= x_n^2 - 5y_n^2 + 4. \end{aligned}$$

(b) Let a and b be positive integers such that $a^2 - 5b^2 + 4 = 0$. Clearly, $a = b = 1$ if $a = 1$ or $b = 1$. Also, we see that a and b have the same parity. Now, suppose $a > 1$ and $b > 1$. Let $\begin{pmatrix} a_1 \\ b_1 \end{pmatrix} = A^{-1}\begin{pmatrix} a \\ b \end{pmatrix}$; that is,

$$a_1 = \frac{3a - 5b}{2} \quad \text{and} \quad b_1 = \frac{-a + 3b}{2}.$$

Then a_1 and b_1 are integers, and $a_1^2 - 5b_1^2 + 4 = a^2 - 5b^2 + 4 = 0$. We have $a - a_1 = \frac{5b - a}{2}$ and $b - b_1 = \frac{a - b}{2}$. The following calculations show that a_1, b_1, $a - a_1$, and $b - b_1$ are all positive:

$$\begin{aligned} (3a - 5b)(a + 3b) &= 4(ab - 3) > 0, \\ (3b - a)(3a + 5b) &= 4(ab + 3) > 0, \\ (5b - a)(5b + a) &= 25b^2 - a^2 = 4(a^2 + 5) > 0, \\ (a - b)(a + b) &= a^2 - b^2 = 4(b^2 - 1) > 0. \end{aligned}$$

If $a_1 = 1$ or $b_1 = 1$, then $a_1 = b_1 = 1 = x_0 = y_0$ and

$$\begin{pmatrix} a \\ b \end{pmatrix} = A \begin{pmatrix} x_0 \\ y_0 \end{pmatrix} = \begin{pmatrix} x_1 \\ y_1 \end{pmatrix}.$$

Otherwise, we iterate the process until we get $a_k = 1 = b_k$ (which will necessarily occur, since we have decreasing sequences of positive integers). Then

$$\begin{pmatrix} a \\ b \end{pmatrix} = A^k \begin{pmatrix} a_k \\ b_k \end{pmatrix} = A^k \begin{pmatrix} x_0 \\ y_0 \end{pmatrix} = \begin{pmatrix} x_k \\ y_k \end{pmatrix};$$

that is, $a = x_k$ and $b = y_k$.

Prove that

$$(a + 3b)(b + 4c)(c + 2a) \geq 60abc$$

for all real numbers $0 \leq a \leq b \leq c$.

Solution

The AM–GM Inequality gives us

$$\frac{a + 3b}{4} \geq a^{\frac{1}{4}} b^{\frac{3}{4}}, \qquad \frac{b + 4c}{5} \geq b^{\frac{1}{5}} c^{\frac{4}{5}}, \qquad \frac{c + 2a}{3} \geq c^{\frac{1}{3}} a^{\frac{2}{3}}.$$

These three inequalities imply that

$$(a + 3b)(b + 4c)(c + 2a) \geq 60 a^{\frac{11}{12}} b^{\frac{19}{20}} c^{\frac{17}{15}}.$$

Now, since $0 \leq a \leq b \leq c$, we have

$$c^{\frac{17}{15}} = c^{\frac{1}{12}} c^{\frac{1}{20}} c \geq a^{\frac{1}{12}} b^{\frac{1}{20}} c,$$

and hence, $60 a^{\frac{11}{12}} b^{\frac{19}{20}} c^{\frac{17}{15}} \geq 60abc$. This proves the proposed inequality.

Determine all positive integers x, n satisfying the equation $x^3 + 3367 = 2^n$.

Solution

One solution is $x = 9$, $n = 12$, since $9^3 + 3367 = 4096 = 2^{12}$. We will prove that this is the only solution.

Let x, n satisfy the given equation. Since $3367 = 7 \times 13 \times 37$, we have $x^3 \equiv 2^n \pmod{7}$. This implies that n is a multiple of 3, since otherwise 2^n is congruent to 2 or 4 modulo 7, while the cube of an integer is congruent to 0, 1, or 6 modulo 7. Thus, $n = 3m$ for some positive integer m. The equation becomes $(2^m)^3 - x^3 = 3367$; that is,

$$(2^m - x)((2^m - x)^2 + 3x \cdot 2^m) = 3367. \tag{1}$$

It follows that $d = 2^m - x$ is a divisor of 3367 such that $d^3 < 3367$. The only possibilities are $d = 1$, $d = 7$, or $d = 13$. The case $d = 1$ transforms (1) into $2^m(2^m - 1) = 2 \times 561$, which clearly has no solution, and the case $d = 13$ transforms (1) into $2^m(2^m - 13) = 2 \times 15$, which also has no solution. That leaves $d = 7$, which yields $2^m(2^m - 7) = 2^4 \times 3^2$. Hence, $m = 4$. As a result, $n = 3m = 12$ and $x = 9$.

Find all solutions $(x, y, z) \in \mathbb{R} \times \mathbb{R} \times \mathbb{R}$ of the system

$$\frac{4x^2}{1 + 4x^2} = y, \qquad \frac{4y^2}{1 + 4y^2} = z, \qquad \frac{4z^2}{1 + 4z^2} = x.$$

Solution

Clearly, x, y, and z must be non-negative, and they are all zero if any one of them is zero. Assume, then, that x, y, and z are positive.

By the AM–GM Inequality, we have $1 + 4x^2 \geq 2\sqrt{4x^2} = 4x$ and hence, $y \leq \dfrac{4x^2}{4x} = x$. Similarly, $z \leq \dfrac{4y^2}{4y} = y$ and $x \leq \dfrac{4z^2}{z} = z$. It follows that $x = y = z$. Setting $y = x$ in the equation $\dfrac{4x^2}{1 + 4x^2} = y$, we obtain $1 + 4x^2 = 4x$; that is, $(2x - 1)^2 = 0$, or $x = \frac{1}{2}$.

Thus, there are exactly two solutions, namely $(x, y, z) = (0, 0, 0)$ and $(x, y, z) = (\frac{1}{2}, \frac{1}{2}, \frac{1}{2})$.

Find all integers $n \geq 2$ for which the system of equations

$$\begin{cases} x_1^2 + x_2^2 + 50 &= 16x_1 + 12x_2 \\ x_2^2 + x_3^2 + 50 &= 16x_2 + 12x_3 \\ x_3^2 + x_4^2 + 50 &= 16x_3 + 12x_4 \\ \cdots\cdots\cdots\cdots \quad \cdots \quad \cdots\cdots \\ x_{n-1}^2 + x_n^2 + 50 &= 16x_{n-1} + 12x_n \\ x_n^2 + x_1^2 + 50 &= 16x_n + 12x_1 \end{cases}$$

has a solution in integers $x_1, x_2, x_3, \ldots, x_n$.

Solution

The desired integers are those which are divisible by 3.

Let $n \geq 2$ be an integer for which the system has a solution in integers $x_1, x_2, \ldots, x_n$. Subscripts are considered modulo n.

For all i, we have

$$x_i^2 + x_{i+1}^2 + 50 = 16x_i + 12x_{i+1};$$

that is, $(x_i - 8)^2 + (x_{i+1} - 6)^2 = 50$. Since the decompositions of 50 as a sum of two squares are $50 = 1 + 49 = 25 + 25$, we deduce that

$$\begin{array}{rlcl} & x_i - 8 = \pm 1 & \text{and} & x_{i+1} - 6 = \pm 7, \\ \text{or} & x_i - 8 = \pm 7 & \text{and} & x_{i+1} - 6 = \pm 1, \\ \text{or} & x_i - 8 = \pm 5 & \text{and} & x_{i+1} - 6 = \pm 5. \end{array}$$

That is, (x_i, x_{i+1}) is one of the pairs $(9, 13)$, $(7, -1)$, $(9, -1)$, $(7, 13)$, $(15, 7)$, $(1, 5)$, $(15, 5)$, $(1, 7)$, $(13, 11)$, $(3, 1)$, $(13, 1)$, $(1, 11)$.

Each of the x_i's must be the first entry in one of these pairs and the second entry in another. We deduce that $x_i \in \{1, 7, 13\}$ for all i. Moveover, if $x_i = 7$ then $x_{i+1} = 13$, $x_{i+2} = 1$, $x_{i+3} = 7$, and so on. It follows that $x_i = x_j$ if and only if $i \equiv j \pmod 3$. Since $x_i = x_{n+i}$, we deduce that $n \equiv 0 \pmod 3$.

Conversely, if $n = 3k$ for some integer $k \geq 1$, we let $x_{3i+1} = 7$, $x_{3i+2} = 13$, and $x_{3i+3} = 1$, for $i = 0, 1, \ldots, k - 1$. Then the x_i's clearly form a solution of the system in integers.

Does there exist a positive integer n such that

$$27^n + 84^n + 110^n + 133^n = 144^n ?$$

Solution

Let

$$f(x) = \left(\frac{27}{144}\right)^x + \left(\frac{84}{144}\right)^x + \left(\frac{110}{144}\right)^x + \left(\frac{133}{144}\right)^x$$

defined on $(0, \infty)$. It is easy to see that the function f is decreasing. Then the equation $f(x) = 1$ has at most one solution in positive real numbers. Since $f(5) = 1$, it follows that $n = 5$ is the only solution of the problem.

Let a, b and c be real numbers with sum 0. Prove that

$$\frac{a^7 + b^7 + c^7}{7} = \left(\frac{a^5 + b^5 + c^5}{5}\right)\left(\frac{a^2 + b^2 + c^2}{2}\right).$$

Solution

Let a, b, c be the roots of $x^3 + qx - r = 0$, and let $S_k = a^k + b^k + c^k$.
Then $S_1 = a + b + c = 0$ and

$$\begin{aligned} S_2 &= a^2 + b^2 + c^2 \\ &= (a + b + c)^2 - 2(ab + bc + ca) = -2q \end{aligned}$$

Now we have

$$\begin{aligned} S_3 + qS_1 - 3r &= 0 & \implies & & S_3 &= 3r \\ S_4 + qS_2 - rS_1 &= 0 & \implies & & S_4 &= 2q^2 \\ S_5 + qS_3 - rS_2 &= 0 & \implies & & S_5 &= -3qr - 2qr = -5qr \\ S_7 + qS_5 - rS_4 &= 0 & \implies & & S_7 &= 5q^2r + 2q^2r = 7q^2r. \end{aligned}$$

Then $\dfrac{S_7}{7} = q^2r = \dfrac{S_5}{5} \cdot \dfrac{S_2}{2}$.

Prove that

$$\sum_{k=0}^{n} \binom{n}{k}(a+k)^{k-1}(b+n-k)^{n-k-1} = (a+b+n)^{n-1}\left(\frac{1}{a}+\frac{1}{b}\right).$$

Solution

Let S denote the left side of the above equation, and let $N = n - 1$. Using $\binom{n}{k} = \binom{N}{k} + \binom{N}{k-1}$, we write $S = S_1 + S_2$, where

$$S_1 = \sum_{k=0}^{N} \binom{N}{k}(a+k)^{k-1}(b+N+1-k)^{N-k}$$

$$\text{and} \quad S_2 = \sum_{k=1}^{N+1} \binom{N}{k-1}(a+k)^{k-1}(b+N+1-k)^{N-k}.$$

Working with S_1 first, we note that, for each $k = 0, 1, \ldots, N$,

$$\binom{N}{k}(a+k)^{k-1}(b+N+1-k)^{N-k}$$

$$= \binom{N}{k}(a+k)^{k-1}(a+b+N+1-(a+k))^{N-k}$$

$$= \binom{N}{k}(a+k)^{k-1}\sum_{j=0}^{N-k}\binom{N-k}{j}(a+b+N+1)^j(-(a+k))^{N-k-j}$$

$$= \sum_{j=0}^{N-k}\binom{N}{k}\binom{N-k}{j}(-1)^{N-k-j}(a+b+N+1)^j(a+k)^{N-1-j}$$

$$= \sum_{j=0}^{N-k}\binom{N}{j}\binom{N-j}{k}(-1)^{N-k-j}(a+b+N+1)^j(a+k)^{N-1-j}.$$

Summing over k, and interchanging the order of summation, we get

$$S_1 = \sum_{j=0}^{N}\binom{N}{j}(a+b+N+1)^j\sum_{k=0}^{N-j}\binom{N-j}{k}(-1)^{N-k-j}(a+k)^{N-1-j}.$$

Now we make use of the following general result: for every polynomial P with degree $< m$,

$$\sum_{k=0}^{m}(-1)^{m-k}\binom{m}{k}P(k) = 0.$$

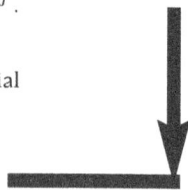

(The sum on the left side is $\Delta^m P(0)$, where Δ is the difference operator defined by $\Delta P(x) = P(x+1) - P(x)$.) It follows that the inner sum in the above expression for S_1 is 0 except when $j = N$. Thus,

$$S_1 = \binom{N}{N}(a+b+N+1)^N \binom{0}{0}(-1)^0(a+0)^{-1} = \frac{1}{a}(a+b+n)^{n-1}.$$

For S_2, we have

$$\begin{aligned} S_2 &= \sum_{k=1}^{N+1} \binom{N}{N-k+1}(a+k)^{k-1}(b+N+1-k)^{N-k} \\ &= \sum_{j=0}^{N} \binom{N}{j}(a+N+1-j)^{N-j}(b+j)^{j-1} \end{aligned}$$

Applying the same argument as for S_1, we get $S_2 = \frac{1}{b}(a+b+n)^{n-1}$. The desired result follows.

For real numbers $x_1, x_2, \ldots, x_6 \in [0, 1]$ prove the inequality

$$\frac{x_1^3}{x_2^5 + x_3^5 + x_4^5 + x_5^5 + x_6^5 + 5} + \frac{x_2^3}{x_1^5 + x_3^5 + x_4^5 + x_5^5 + x_6^5 + 5} + \cdots$$
$$+ \frac{x_6^3}{x_1^5 + x_2^5 + x_3^5 + x_4^5 + x_5^5 + 5} \leq \frac{3}{5}.$$

Solution

Since $x_1, \ldots, x_6$ are in the interval $[0, 1]$,

$$x_2^5 + x_3^5 + x_4^5 + x_5^5 + x_6^5 + 5 \geq x_1^5 + x_2^5 + x_3^5 + x_4^5 + x_5^5 + x_6^5 + 4.$$

By permuting the subscripts, we see that the left side of the inequality in the problem is at most

$$\sum_{i=1}^{6} \frac{x_i^3}{x_1^5 + x_2^5 + \cdots + x_6^5 + 4} = \frac{\sum\limits_{i=1}^{6} x_i^3}{\sum\limits_{i=1}^{6} x_i^5 + 4}.$$

For any $y \geq 0$, the AM–GM Inequality gives us

$$\frac{y^5 + y^5 + y^5 + 1 + 1}{5} \geq \sqrt[5]{y^5 \cdot y^5 \cdot y^5} = y^3;$$

that is, $3y^5 + 2 \geq 5y^3$. Thus,

$$5 \sum_{i=1}^{6} x_i^3 \leq \sum_{i=1}^{6} (3x_i^5 + 2) = 3 \left(\sum_{i=1}^{6} x_i^5 + 4 \right);$$

that is,

$$\frac{\sum\limits_{i=1}^{6} x_i^3}{\sum\limits_{i=1}^{6} x_i^5 + 4} \leq \frac{3}{5}.$$

This, together with our initial observations, completes the proof.

Prove that for all strictly positive numbers a, b, and c the inequality

$$(a + b + x)^{-1} + (b + c + x)^{-1} + (c + a + x)^{-1} \le x^{-1},$$

holds, where $x = \sqrt[3]{abc}$.

Solution

Let $(a + b + x)^{-1} + (b + c + x)^{-1} + (c + a + x)^{-1} = N/D$. An easy calculation yields

$$
\begin{aligned}
N &= 3x^2 + 4x(a + b + c) + a^2 + b^2 + c^2 + 3(ab + bc + ca)\,, \\
D &= x^3 + 2x^2(a + b + c) + x\Big(a^2 + b^2 + c^2 + 3(ab + bc + ca)\Big) \\
&\quad + a^2b + ab^2 + b^2c + bc^2 + c^2a + ca^2 + 2abc\,.
\end{aligned}
$$

Observing that

$$a^2b + ab^2 + b^2c + bc^2 + c^2a + ca^2 = (a + b + c)(ab + bc + ca) - 3abc$$

and $x^3 = abc$, we have

$$
\begin{aligned}
D &= 2x^2(a + b + c) + x\Big(a^2 + b^2 + c^2 + 3(ab + bc + ca)\Big) \\
&\quad + (a + b + c)(ab + bc + ca)\,.
\end{aligned}
$$

Hence, $D - xN = (a + b + c)((ab + bc + ca) - 2x^2) - 3x^3$.

By the AM–GM Inequality, we have $a + b + c \ge 3\sqrt[3]{abc} = 3x$ and $ab + bc + ca \ge 3\sqrt[3]{a^2b^2c^2} = 3x^2$. Thus,

$$D - xN \ge 3x(3x^2 - 2x^2) - 3x^3 = 0\,.$$

It follows that $\dfrac{N}{D} \le \dfrac{1}{x}$, as required.

Find all the integer values of m, for which the equation

$$\left\lfloor \frac{m^2 x - 13}{1999} \right\rfloor = \frac{x - 12}{2000}$$

has 1999 distinct real solutions ($\lfloor \cdot \rfloor$ denotes the integral part function).

Solution

If x is a solution of the given equation, then $x = 2000k + 12$ for some integer k. Moreover, the given equation may be rewritten as $k = m^2 k + \left\lfloor \frac{m^2 k + 12m^2 - 13}{1999} \right\rfloor$; that is, $f(k) = 0$, where

$$f(k) = (m^2 - 1)k + \left\lfloor \frac{m^2 k + 12m^2 - 13}{1999} \right\rfloor .$$

Suppose that $m^2 > 1$. Since f is the sum of an increasing function and a non-decreasing function on $\mathbb{Z}$, the function f is increasing on $\mathbb{Z}$. It follows that the equation $f(k) = 0$ has at most one solution. Therefore, m is not a solution of the problem.

If $m = 0$, then the given equation becomes $-1 = \frac{x - 12}{2000}$, which does not have 1999 solutions. Thus, $m = 0$ is not a solution of the problem.

Finally, suppose that $m^2 = 1$. Then $f(k) = \left\lfloor \frac{k - 1}{1999} \right\rfloor = 0$, and the equation $f(k) = 0$ has solutions $k = 1, 2, \ldots, 1999$.

Thus, the desired values of m are $m = 1$ and $m = -1$.

Prove that for each prime number p the equation

$$2^p + 3^p = a^n$$

has no solutions (a, n), with a and n integers > 1.

Solution

Let $f(p) = 2^p + 3^p$. Note first that $f(2) = 13$ and $f(5) = 275$, neither of which is an nth power for $n > 1$. Now we assume that $f(p) = a^n$ for some integers a and $n > 1$, where p is a prime such that $p \neq 2$ and $p \neq 5$. Since p is odd, we have $f(p) = (2 + 3)(2^{p-1} - 2^{p-2} \cdot 3 + \cdots + 3^{p-1}) = 5A$ where

$$A = \sum_{k=0}^{p-1} 2^{p-k-1}(-3)^k.$$

In particular, $5 \mid f(p)$, and hence, $5 \mid a$. Since $f(p) = a^n$ and $n > 1$, we have $25 \mid f(p)$, and thus, $5 \mid A$. However, since $-3 \equiv 2 \pmod 5$, we get

$$A \equiv \sum_{k=0}^{p-1} 2^{p-k-1} \cdot 2^k \equiv p \cdot 2^{p-1} \pmod 5,$$

which is not divisible by 5, since $5 \nmid p$. This is a contradiction.

(a) Determine all pairs (x, k) of positive integers which satisfy the equation

$$3^k - 1 = x^3.$$

(b) Prove that if n is an integer greater than 1 and different from 3 there are no pairs (x, k) of positive integers satisfying the equation

$$3^k - 1 = x^n.$$

Solution

(a) Clearly, $(x, k) = (2, 2)$ is a solution. We will show that there are no other solutions.

Let x and k be positive integers such that $3^k - 1 = x^3$. Then x is even, since $3^k - 1$ is even, and thus, $x \geq 2$. Furthermore, we have

$$3^k = x^3 + 1 = (x + 1)(x^2 - x + 1).$$

It follows that $x^2 - x + 1 = 3^s$ for some integer $s \geq 1$. (We cannot have $s = 0$ because $x \geq 2$.) However, $x^2 - x + 1$ is not congruent to 0 modulo 9, as we can check by considering $x \in \{0, 1, \ldots, 8\}$. Therefore, $s = 1$. Then $x^2 - x + 1 = 3$, giving $x = 2$, from which $k = 2$ follows.

(b) Let n be an integer greater than 1 for which there exist positive integers x, k such that $3^k - 1 = x^n$. We will prove that $n = 3$.

First, we observe that n cannot be even. For, if $n = 2r$, we would have $(x^r)^2 = 3^k - 1 \equiv -1 \pmod 3$. This is impossible, since a square is congruent to either 0 or 1 modulo 3.

Now set $n = 2r + 1$. Then

$$3^k = x^n + 1 = x^{2r+1} + 1 = (x + 1)a,$$

where $a = 1 - x + x^2 - \cdots + x^{2r}$. Note that $x + 1 > 1$ and also $a > 1$ (because $n > 1$). It follows that $x + 1 = 3^u$ and $a = 3^v$ for some positive integers u and v. Then $x \equiv -1 \pmod 3$ and

$$3^v = a \equiv 1 + 1 + \cdots + 1 \equiv 2r + 1 \pmod 3.$$

Thus, $2r + 1 \equiv 0 \pmod 3$; that is, $r \equiv 1 \pmod 3$. Hence, $r = 1 + 3w$ for some non-negative integer w. Now $3^k = x^{2r+1} + 1 = (x^{1+2w})^3 + 1$. From part (a), we must have $x^{1+2w} = 2$, which implies that $w = 0$. Therefore, $r = 1$ and $n = 3$.

Find all positive integers n for which there are k integers $n_1, n_2, \ldots, n_k$, each greater than 3, such that

$$n = n_1 n_2 \cdots n_k = \sqrt[2^k]{2^{(n_1-1)(n_2-1)\cdots(n_k-1)}} - 1.$$

Solution

Let n be a positive integer, and let $n_1, n_2, \ldots, n_k$ be integers, each greater than 3. Suppose that the equation in the problem is satisfied. Then

$$n = n_1 \cdots n_k = 2^m - 1, \tag{1}$$

where

$$\begin{aligned} m &= \frac{(n_1-1)(n_2-1)\cdots(n_k-1)}{2^k} \\ &= \left(\frac{n_1-1}{2}\right)\left(\frac{n_2-1}{2}\right)\cdots\left(\frac{n_k-1}{2}\right). \end{aligned} \tag{2}$$

From (1), we see that m is a positive integer, since $2^m = n + 1$ is a positive integer. We also see that n is odd and each n_i is odd. Since $n_i > 3$, we must have $n_i \geq 5$, for each i.

Lemma. For any integer $j \geq 10$, we have $2^j - 1 > j^3$.

Proof. We proceed by induction on j. The inequality is true for $j = 10$, since $2^{10} - 1 = 1023 > 10^3 = 1000$. Next, we assume that $2^j - 1 > j^3$ for some fixed integer $j \geq 10$. Let us prove that $2^{j+1} - 1 > (j+1)^3$. Since $j \geq 10$, we have

$$\left(\frac{j+1}{j}\right)^3 < \left(\frac{5}{4}\right)^3 < 2.$$

Thus, $2^{j+1} - 1 > 2^{j+1} - 2 = 2(2^j - 1) > 2j^3 > (j+1)^3$.

Now, let us return to our problem. Suppose that $m \geq 10$. Using the lemma and (2), we have

$$2^m - 1 > m^3 = \left(\frac{n_1-1}{2}\right)^3 \cdots \left(\frac{n_k-1}{2}\right)^3. \tag{3}$$

For each i, since $n_i \geq 5$, we must have

$$\left(\frac{n_i-1}{2}\right)^3 \geq 4 \cdot \frac{n_i-1}{2} > n_i.$$

Using (3), we deduce that $2^m - 1 > n_1 n_2 \cdots n_k$, which contradicts (1).

Thus, there are no solutions with $m \geq 10$. It is easy to check that the only positive integer $m \leq 9$ for which (1) and (2) can be satisfied is $m = 3$, which gives $n = 7$.

Prove that there exist ten different real numbers $a_1, a_2, \ldots, a_{10}$ such that the equation

$$(x - a_1)(x - a_2) \cdots (x - a_{10}) = (x + a_1)(x + a_2) \cdots (x + a_{10})$$

has exactly 5 different real roots.

Solution

Let $a_1, a_2, \ldots, a_{10}$ be distinct real numbers such that $a_1, \ldots, a_5$ are positive, $a_6 = 0$, $a_7 + a_8 = 0$, and $a_9 + a_{10} = 0$. For $k \in \{6, 7, 8, 9, 10\}$, the factor $x - a_k$ occurs on both sides of the given equation and, hence, a_k is a real root of the equation.

For $x \notin \{a_6, a_7, a_8, a_9, a_{10}\}$, the equation reduces to

$$(x - a_1)(x - a_2) \ldots (x - a_5) = (x + a_1)(x + a_2) \ldots (x + a_5). \qquad (1)$$

If $x > 0$, then, for each $k \in \{1, 2, 3, 4, 5\}$,

$$|x - a_k| = \max\{x - a_k, a_k - x\} < x + a_k = |x + a_k|,$$

and (1) cannot hold. By symmetry, (1) cannot hold if $x < 0$. Neither does (1) hold if $x = 0$. Therefore, (1) has no real roots. Hence, the equation in the problem statement has no real roots besides $a_6, a_7, a_8, a_9, a_{10}$.

Let $a_1, a_2, \ldots, a_{2000}$ be real numbers such that

$$a_1^3 + a_2^3 + \cdots + a_n^3 = (a_1 + a_2 + \cdots + a_n)^2$$

for all n, $1 \le n \le 2000$. Prove that every element of the sequence is an integer.

Solution

First we note that if $a_i = i$ for $i = 1, 2, \ldots, n$, then

$$a_1^3 + a_2^3 + \cdots + a_n^3 = \sum_{i=1}^{n} i^3 = \left(\frac{n(n+1)}{2} \right)^2$$

$$= \left(\sum_{i=1}^{n} i \right)^2 = (a_1 + a_2 + \cdots + a_n)^2.$$

Lemma. If $a_i = i$ for $i = 1, 2, \ldots, n$ and $\sum_{i=1}^{n+1} a_i^3 = \left(\sum_{i=1}^{n+1} a_i \right)^2$, then $a_{n+1} \in \{n+1, -n, 0\}$.

Proof: Starting with the given relation $\sum_{i=1}^{n+1} a_i^3 = \left(\sum_{i=1}^{n+1} a_i \right)^2$, we get

$$a_{n+1}^3 + \sum_{i=1}^{n} i^3 = \left(a_{n+1} + \sum_{i=1}^{n} i \right)^2$$

$$= a_{n+1}^2 + 2a_{n+1} \left(\frac{n(n+1)}{2} \right) + \left(\sum_{i=1}^{n} i \right)^2,$$

which simplifies to $a_{n+1}(a_{n+1} + n)(a_{n+1} - (n+1)) = 0$. The conclusion of the lemma follows.

For each positive integer k, let $\mathcal{P}_k$ be the claim:

If $a_1, a_2, \ldots, a_k$ are real numbers such that $\sum\limits_{i=1}^{n} a_i^3 = \left(\sum\limits_{i=1}^{n} a_i \right)^2$ for all integers n such that $1 \le n \le k$, then $a_1, a_2, \ldots, a_k$ are integers.

We will prove, by induction on k, that $\mathcal{P}_k$ holds for each k. First note that $\mathcal{P}_1$ holds, since the equation $a_1^3 = a_1^2$ implies that $a_1 \in \{0, 1\}$. Let $k \ge 1$ be a fixed integer, and suppose that $\mathcal{P}_i$ holds for $i = 1, 2, \ldots, k$. Let $a_1, a_2, \ldots, a_{k+1}$ be real numbers such that $\sum\limits_{i=1}^{n} a_i^3 = \left(\sum\limits_{i=1}^{n} a_i \right)^2$ for all integers n such that $1 \le n \le k+1$.

Case 1. There exists $i \in \{1, 2, \ldots, k+1\}$ such that $a_i = 0$.

Delete a_i from the sequence $a_1, a_2, \ldots, a_{k+1}$. The remaining k numbers in the sequence satisfy the hypothesis of $\mathcal{P}_k$. Since $\mathcal{P}_k$ holds, each of these numbers must be an integer. Thus, all of $a_1, a_2, \ldots, a_{k+1}$ are integers.

Case 2. There exists $i \in \{1, 2, \ldots, k+1\}$ such that $a_{i+1} = -a_i$.

Then $k \ge 2$. The fact that $\mathcal{P}_i$ holds implies that $a_1, a_2, \ldots, a_i$ are integers. Then $a_{i+1} = -a_i$ is an integer. Now delete a_i and a_{i+1} from the sequence $a_1, a_2, \ldots, a_{k+1}$. The remaining $k - 1$ numbers in the sequence satisfy the hypothesis of $\mathcal{P}_{k-1}$. Since $\mathcal{P}_{k-1}$ holds, each of these numbers must be an integer. Thus, all of $a_1, a_2, \ldots, a_{k+1}$ are integers.

Case 3. For each $i \le k+1$, we have $a_i \ne 0$ and $a_i \ne -a_{i-1}$.

Then an easy induction, using the lemma above, shows that $a_i = i$ for all $i \in \{1, 2, \ldots, k+1\}$. Thus, all of $a_1, a_2, \ldots, a_{k+1}$ are integers.

This ends the induction step, and we are done.

Prove that

$$\frac{1}{\sqrt{1+x^2}} + \frac{1}{\sqrt{1+y^2}} < x, y \leq 1.$$

Solution

The problem should read: prove that

$$\frac{1}{\sqrt{1+x^2}} + \frac{1}{\sqrt{1+y^2}} \leq \frac{2}{\sqrt{1+xy}} \quad \text{for } 0 \leq x, y \leq 1.$$

If $x = 0$, then the inequality reduces to $1 + \dfrac{1}{\sqrt{1+y^2}} \leq 2$, which is true, since $y > 0$. By symmetry, the inequality is also true if $y = 0$.

Now, suppose that $0 < x \leq 1$ and $0 < y \leq 1$. Let $u \geq 0$ and $v \geq 0$ be such that $x = e^{-u}$ and $y = e^{-v}$. Then the inequality becomes

$$\frac{1}{\sqrt{1+e^{-2u}}} + \frac{1}{\sqrt{1+e^{-2v}}} \leq \frac{2}{\sqrt{1+e^{-(u+v)}}} \, ;$$

that is,

$$\frac{f(u) + f(v)}{2} \leq f\left(\frac{u+v}{2}\right) ,$$

where $f(x) = \dfrac{1}{\sqrt{1+e^{-2x}}}$. Since $f''(x) = \dfrac{1 - 2e^{2x}}{(1 + e^{-2x})^{5/2}e^{4x}}$, the function f is concave on an interval containing $[0, \infty)$. Therefore, the above inequality is true.

Prove that for any prime p, there exist integers x, y, z, and w such that $x^2 + y^2 + z^2 - wp = 0$ and $0 < w < p$.

Solution

For $p = 2$, we set $x = y = w = 1$ and $z = 0$.

For p odd, we note that the numbers

$$1 + 0^2, \quad 1 + 1^2, \quad \ldots, \quad 1 + \left(\frac{p-1}{2}\right)^2$$

leave $(p+1)/2$ distinct remainders upon division by p, as do the numbers

$$-0^2, \quad -1^2, \quad \ldots, \quad -\left(\frac{p-1}{2}\right)^2 .$$

Indeed, if $0 \leq a, b \leq (p-1)/2$, then p divides $a^2 - b^2 = (a-b)(a+b)$ if and only if p divides $a + b$ or $a - b$, which is impossible if $a \neq b$.

Thus, there are at least two integers $0 \leq x, y \leq (p-1)/2$ such that

$$1 + x^2 \equiv -y^2 \pmod{p} .$$

Hence, $x^2 + y^2 + 1 = wp$ for some integer w. Since

$$0 < x^2 + y^2 + 1 \leq 2\left(\frac{p-1}{2}\right)^2 + 1 < p^2 ,$$

we have $0 < w < p$, as desired.

The real numbers a, b, c, x, y, and z are such that $a > b > c > 0$ and $x > y > z > 0$. Prove that

$$\frac{a^2x^2}{(by+cz)(bz+cy)} + \frac{b^2y^2}{(cz+ax)(cx+az)} + \frac{c^2z^2}{(ax+by)(ay+bx)} \geq \frac{3}{4}.$$

Solution

We will prove the given inequality under the slightly weaker hypothesis that $a \geq b \geq c > 0$ and $x \geq y \geq z > 0$. For convenience, we let S denote the expression on the left side of the inequality.

From $a \geq b \geq c > 0$ and $x \geq y \geq z > 0$, we deduce that

$$a^2x^2 \geq b^2y^2 \geq c^2z^2. \tag{1}$$

Moreover,

$$0 < by + cz \leq cz + ax \leq ax + by$$

and

$$0 < bz + cy \leq cx + az \leq ay + bx.$$

Then

$$\begin{aligned}
\frac{1}{(by+cz)(bz+cy)} &\geq \frac{1}{(cz+ax)(cx+az)} \\
&\geq \frac{1}{(ax+by)(ay+bx)} > 0. \tag{2}
\end{aligned}$$

From (1), (2), and Chebyshev's Inequality, it follows that

$$S \geq \tfrac{1}{3}(a^2x^2 + b^2y^2 + c^2z^2) \cdot$$
$$\cdot \left(\frac{1}{(by+cz)(bz+cy)} + \frac{1}{(cz+ax)(cx+az)} + \frac{1}{(ax+by)(ay+bx)} \right).$$

Using the AM–HM Inequality, we get

$$\frac{1}{3} \left(\frac{1}{(by+cz)(bz+cy)} + \frac{1}{(cz+ax)(cx+az)} + \frac{1}{(ax+by)(ay+bx)} \right)$$
$$\geq \frac{3}{(by+cz)(bz+cy) + (cz+ax)(cx+az) + (ax+by)(ay+bx)}.$$

Hence,

$$S \geq \frac{3(a^2x^2 + b^2y^2 + c^2z^2)}{S'}, \tag{3}$$

where

$$\begin{aligned}
S' &= (by+cz)(bz+cy) + (cz+ax)(cx+az) + (ax+by)(ay+bx) \\
&= a^2(xy+xz) + b^2(yz+yx) + c^2(zx+zy) \\
&\quad + (ab+ac)x^2 + (bc+ba)y^2 + (ca+cb)z^2.
\end{aligned}$$

Note that $xy + xz \leq \frac{1}{2}(x^2 + y^2) + \frac{1}{2}(x^2 + z^2) = x^2 + \frac{1}{2}(y^2 + z^2)$, with equality if and only if $x = y = z$. Using this and similar inequalities, we deduce that

$$
\begin{aligned}
S' &\leq a^2\left(x^2 + \tfrac{1}{2}y^2 + \tfrac{1}{2}z^2\right) + b^2\left(\tfrac{1}{2}x^2 + y^2 + \tfrac{1}{2}z^2\right) \\
&\quad + c^2\left(\tfrac{1}{2}x^2 + \tfrac{1}{2}y^2 + z^2\right) + \left(a^2 + \tfrac{1}{2}b^2 + \tfrac{1}{2}c^2\right)x^2 \\
&\quad\quad + \left(\tfrac{1}{2}a^2 + b^2 + \tfrac{1}{2}c^2\right)y^2 + \left(\tfrac{1}{2}a^2 + \tfrac{1}{2}b^2 + c^2\right)z^2 \\
&= (a^2 + b^2 + c^2)(x^2 + y^2 + z^2) + a^2x^2 + b^2y^2 + c^2z^2,
\end{aligned}
$$

with equality if and only if $x = y = z$ and $a = b = c$. Using Chebyshev's Inequality again, we get

$$
\begin{aligned}
S' &\leq 3(a^2x^2 + b^2y^2 + c^2z^2) + a^2x^2 + b^2y^2 + c^2z^2 \\
&= 4(a^2x^2 + b^2y^2 + c^2z^2). \quad\quad\quad\quad\quad\quad\quad\quad (4)
\end{aligned}
$$

From (3) and (4), we deduce that $S \geq \frac{3}{4}$. Equality occurs if and only if $x = y = z$ and $a = b = c$.

Pete and Bill play the following game. At the beginning, Pete chooses a number a, then Bill chooses a number b, and then Pete chooses a number c. Can Pete choose his numbers in such a way that the three equations $x^3 + ax^2 + bx + c = 0$, $x^3 + bx^2 + cx + a = 0$, and $x^3 + cx^2 + ax + b = 0$ have a common

(a) real root?

(b) negative root?

Solution

(a) Yes. Pete simply chooses $a = -1$, and then, for any b chosen by Bill, Pete chooses $c = -b$. The three equations become $x^3 - x^2 + bx - b = 0$, $x^3 + bx^2 - bx - 1 = 0$, and $x^3 - bx^2 - x + b = 0$, which clearly have $x = 1$ as a common root.

(b) No. Suppose Bill chooses $b = 0$, and suppose r is a negative root common to all three equations. Then, in particular, we have

$$r^3 + cr + a = 0 \qquad\qquad (1)$$
$$\text{and} \quad r^3 + cr^2 + ar = 0 . \qquad (2)$$

From (2) we get $r^2 + cr + a = 0$, or $cr + a = -r^2$. Substituting into (1), we then get $r^3 - r^2 = 0$, which yields $r = 0$ or 1, a contradiction.

How many pairs (n, q) satisfy $\{q^2\} = \left\{\dfrac{n!}{2000}\right\}$, where n is a positive integer and q is a non-integer rational number such that $0 < q < 2000$? $\{r\}$ means the "fractional part" of r.

Solution

There are 2400 such pairs.

Let n be a positive integer and q a non-integer rational number such that $0 < q < 2000$. Let a and b be positive integers such that $q = a/b$ and $\gcd(a, b) = 1$. Then $b > 1$ (since q is not an integer). We say that (n, q) is a *good pair* if $\{q^2\} = \left\{\dfrac{n!}{2000}\right\}$. The problem is to find how many pairs (n, q) are good pairs.

For any real number x, let $\lfloor x \rfloor = x - \{x\}$ (the integer part of x). Then (n, q) is a good pair if and only if

$$\frac{a^2}{b^2} - \left\lfloor \frac{a^2}{b^2} \right\rfloor = \frac{n!}{2000} - \left\lfloor \frac{n!}{2000} \right\rfloor ; \qquad (1)$$

that is,

$$2000a^2 = b^2 \left(2000\lfloor q^2 \rfloor + n! - 2000 \left\lfloor \frac{n!}{2000} \right\rfloor \right) .$$

If this condition is satisfied, then, according to Gauss' Theorem, b^2 divides 2000, and it follows that $b \in \{2, 4, 5, 10, 20\}$.

Suppose that $b \neq 5$. Then $b = 2\tilde{b}$, where $\tilde{b} \in \{1, 2, 5, 10\}$, and a is odd, since $\gcd(a, b) = 1$. From (1), we have

$$a^2 = 4\tilde{b}^2 \lfloor q^2 \rfloor + \frac{n!\tilde{b}^2}{500} - 4\tilde{b}^2 \left\lfloor \frac{n!}{2000} \right\rfloor .$$

It follows that $\dfrac{n!\tilde{b}^2}{500}$ is an odd integer. This cannot be true if $n!$ is divisible by 8; therefore, $n \leq 3$. But then $\dfrac{n!\tilde{b}^2}{500}$ is not an integer, a contradiction.

Thus, we must have $b = 5$. Then a is not divisible by 5. From (1), we have

$$a^2 = 25\lfloor q^2 \rfloor + \frac{n!}{80} - 25 \left\lfloor \frac{n!}{2000} \right\rfloor .$$

It follows that $\dfrac{n!}{80}$ is an integer not divisible by 5; that is, $n \in \{6, 7, 8, 9\}$.

We have reduced the problem to that of finding good pairs (n, q) of the form $\left(n, \dfrac{a}{5}\right)$, where $n \in \{6, 7, 8, 9\}$ and a is a positive integer not divisible by 5. Moreover, note that for any positive integer a,

$$\left\{ \left(\frac{25 \pm a}{5} \right)^2 \right\} = \left\{ 25 \pm 2a + \frac{a^2}{25} \right\} = \left\{ \left(\frac{a}{5} \right)^2 \right\} .$$

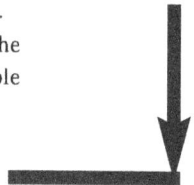

Therefore, if $\left(n, \frac{a}{5}\right)$ is a good pair, then $\left(n, \frac{25+a}{5}\right)$ is a good pair, and so is $\left(n, \frac{25-a}{5}\right)$ if $0 < a < 25$. Hence, we will be able to determine all good pairs $\left(n, \frac{a}{5}\right)$ by finding those pairs for which $1 \leq a \leq 12$ (and a is not divisible by 5).

Case 1. $n = 6$.

Then $\left\{\frac{n!}{2000}\right\} = \frac{9}{25}$, and $\left(6, \frac{a}{5}\right)$ is a good pair if and only if $\left\{\frac{a^2}{25}\right\} = \frac{9}{25}$. It is easy to check that $a = 3$ is the only solution satisfying $1 \leq a \leq 12$. It follows that the good pairs of the form $(6, q)$ are all pairs of the form $\left(6, \frac{3+25k}{5}\right)$ and $\left(6, \frac{22+25k}{5}\right)$, for $k = 0, 1, \ldots, 399$. There are 800 such good pairs.

Case 2. $n = 7$.

Then $\left\{\frac{n!}{2000}\right\} = \frac{13}{25}$, and $\left(7, \frac{a}{5}\right)$ is a good pair if and only if $\left\{\frac{a^2}{25}\right\} = \frac{13}{25}$. If this equation is satisfied, then

$$a^2 = 25 \left\lfloor \frac{a^2}{25} \right\rfloor + 13 \equiv 3 \pmod 5 .$$

But a square is never congruent to 3 (mod 5). Thus, there is no good pair in this case.

Case 3. $n = 8$.

Then $\left\{\frac{n!}{2000}\right\} = \frac{4}{25}$, and $\left(8, \frac{a}{5}\right)$ is a good pair if and only if $\left\{\frac{a^2}{25}\right\} = \frac{4}{25}$. Here $a = 2$ is the only solution satisfying $1 \leq a \leq 12$. It follows that the good pairs of the form $(8, q)$ are all pairs of the form $\left(8, \frac{2+25k}{5}\right)$ and $\left(8, \frac{23+25k}{5}\right)$, for $k = 0, 1, \ldots, 399$. There are 800 such good pairs.

Case 4. $n = 9$.

Then $\left\{\frac{n!}{2000}\right\} = \frac{11}{25}$, and $\left(9, \frac{a}{5}\right)$ is a good pair if and only if $\left\{\frac{a^2}{25}\right\} = \frac{11}{25}$. Now $a = 6$ is the only solution satisfying $1 \leq a \leq 12$. It follows that the good pairs of the form $(9, q)$ are all pairs of the form $\left(9, \frac{6+25k}{5}\right)$ and $\left(9, \frac{19+25k}{5}\right)$, for $k = 0, 1, \ldots, 399$. There are 800 such good pairs, and we are done.

(a) Find all positive integers n such that $(a^a)^n = b^b$ has at least one solution in integers a and b, both exceeding 1.

(b) Find all positive integers a and b such that $(a^a)^5 = b^b$.

Solution

(a) All positive integers n except $n = 2$ are solutions.

If $n = 1$, just choose $a = b \geq 2$.

If $n \geq 3$, choose $a = (n-1)^{n-1}$ and $b = (n-1)^n$. Then $a, b \geq 2$, and

$$(a^a)^n = a^{an} = (n-1)^{n(n-1)^n} = ((n-1)^n)^{(n-1)^n} = b^b.$$

Now suppose that there exist integers $a, b \geq 2$ such that

$$(a^a)^2 = b^b. \tag{1}$$

We cannot have $b \leq a$, because this gives $b^b \leq a^a < (a^a)^2$, since $a > 1$. We cannot have $2a \leq b$, because then $b^b \geq (2a)^{2a} = 2^{2a}(a^a)^2 > (a^a)^2$. Thus, we must have $a < b < 2a$. It follows that a does not divide b.

Let p be a prime number which divides a. Then, from (1), p divides b. Let α and β be the exponents of p in the prime decomposition of a and b, respectively. From (1), we have $2a\alpha = b\beta$. Then $\dfrac{\alpha}{\beta} = \dfrac{b}{2a} < 1$. Thus, $\alpha < \beta$. Since this is true for each prime p dividing a, it follows that a divides b, a contradiction.

We conclude that $n = 2$ is not a solution.

(b) Clearly, $(a, b) = (1, 1)$ is a solution. Now suppose that $a > 1$ and b are positive integers such that

$$(a^a)^5 = b^b. \tag{2}$$

As in (a), where we proved that $a < b < 2a$, we now deduce that $a < b < 5a$.

Let p be a prime number which divides a. Then from (2), p divides b. Letting α and β be the exponents of p in the prime decomposition of a and b, respectively, we obtain $\alpha < \beta$, as in (a), from which we again deduce that a divides b. Then $b = ka$, where $k \in \{2, 3, 4\}$ (since $1 < a < b < 5a$). From (2), we have $a^{5a} = (ka)^{ka}$. Thus, $a^5 = (ka)^k$, which leads to $a^{5-k} = k^k$.

If $k = 2$, we must have $a^3 = 4$, which is impossible. If $k = 3$, then we need $a^2 = 27$, which is again impossible. If $k = 4$, our equation becomes $a = 4^4 = 256$, which leads to $b = 4^5 = 1024$. Conversely, we have seen in (a) that $(4^4, 4^5)$ is a solution of (2).

Thus, the solutions of (2) are $(1, 1)$ and $(256, 1024)$.

Find all pairs (x, y) of positive integers such that $y^{x^2} = x^{y+2}$.

Solution

Note that $(1, 1)$ and $(2, 2)$ are solutions. We prove that these are the only solutions.

Let x and y be positive integers such that

$$y^{x^2} = x^{y+2}. \tag{1}$$

If either $x = 1$ or $y = 1$, then $x = y = 1$. Now assume that $x > 1$ and $y > 1$. From (1), we see that x and y have exactly the same prime divisors. Let $x = \prod p_i^{\alpha_i}$ and $y = \prod p_i^{\beta_i}$ be the prime decompositions of x and y. Then, from (1), for each j,

$$x^2 \beta_j = (y + 2)\alpha_j. \tag{2}$$

Now suppose that one of the prime divisors of x and y, say p_i, is odd. Then $\gcd(p_i, y + 2) = 1$. Since $p_i^{\alpha_i}$ divides x, we see that $p_i^{2\alpha_i}$ divides x^2. From (2), we deduce that $p_i^{2\alpha_i}$ divides α_j for each j. In particular, $p_i^{2\alpha_i}$ divides α_i. It follows that $9^{\alpha_i} < p_i^{2\alpha_i} \leq \alpha_i$. But an easy induction shows that $9^n > n$ for each positive integer n. We have arrived at a contradiction.

Thus, the unique prime divisor of x and y is 2. Let $x = 2^\alpha$ and $y = 2^\beta$, with $\alpha, \beta \geq 1$. Equation (1) reduces to $2^{2\alpha-1}\beta = \alpha(2^{\beta-1} + 1)$. If $\beta \geq 2$, then, since $2^{\beta-1} + 1$ is odd, we see that $2^{2\alpha-1}$ divides α. But this is not possible, since $2^{2n-1} > n$ for each integer $n \geq 1$ (by an easy induction). Therefore, $\beta = 1$, and the equation reduces to $2^{2\alpha-2} = \alpha$. Since $2^{2n-2} > n$ for each integer $n \geq 2$ (by another easy induction), it follows that $\alpha = 1$. Thus, $x = y = 2$.

Prove that

$$\frac{1}{\sqrt{1+x^2}} + \frac{1}{\sqrt{1+y^2}} \leq \frac{2}{\sqrt{1+xy}} \quad \text{for } 0 < x, y \leq 1.$$

Solution

We prove the inequality under the slightly weaker conditions that $x, y \geq 0$ and $xy \leq 1$.

For $x, y \geq 0$, let

$$F(x,y) = \sqrt{1+xy}\left(\frac{2}{\sqrt{1+xy}} - \frac{1}{\sqrt{1+x^2}} - \frac{1}{\sqrt{1+y^2}}\right).$$

Then

$$
\begin{aligned}
F(x,y) &= \frac{\sqrt{1+x^2} - \sqrt{1+xy}}{\sqrt{1+x^2}} + \frac{\sqrt{1+y^2} - \sqrt{1+xy}}{\sqrt{1+y^2}} \\
&= \frac{x(x-y)}{1+x^2 + \sqrt{1+xy}\sqrt{1+x^2}} + \frac{y(y-x)}{1+y^2 + \sqrt{1+xy}\sqrt{1+y^2}} \\
&= \frac{(x-y)G(x,y)}{\left(1+x^2 + \sqrt{1+xy}\sqrt{1+x^2}\right)\left(1+y^2 + \sqrt{1+xy}\sqrt{1+y^2}\right)},
\end{aligned}
$$

where

$$
\begin{aligned}
G(x,y) &= x\left(1+y^2 + \sqrt{1+xy}\sqrt{1+y^2}\right) \\
&\quad - y\left(1+x^2 + \sqrt{1+xy}\sqrt{1+x^2}\right) \\
&= x - y - xy(x-y) + \sqrt{1+xy}\left(x\sqrt{1+y^2} - y\sqrt{1+x^2}\right) \\
&= x - y - xy(x-y) + \frac{\sqrt{1+xy}(x^2 - y^2)}{x\sqrt{1+y^2} + y\sqrt{1+x^2}} \\
&= (x-y)\left(1 - xy + \frac{\sqrt{1+xy}(x+y)}{x\sqrt{1+y^2} + y\sqrt{1+x^2}}\right).
\end{aligned}
$$

Thus,

$$
F(x,y) = \frac{(x-y)^2\left(1 - xy + \dfrac{\sqrt{1+xy}(x+y)}{x\sqrt{1+y^2} + y\sqrt{1+x^2}}\right)}{\left(1+x^2 + \sqrt{1+xy}\sqrt{1+x^2}\right)\left(1+y^2 + \sqrt{1+xy}\sqrt{1+y^2}\right)}.
$$

It is now evident that $F(x,y) \geq 0$ if $xy \leq 1$. The desired result follows.

Determine all triplets of positive real numbers x, y, and z solving the system of equations

$$x + y + z = 6,$$
$$\frac{1}{x} + \frac{1}{y} + \frac{1}{z} = 2 - \frac{4}{xyz}.$$

Solution

Suppose that x, y, z satisfy the first equation, $x + y + z = 6$. Then, applying the AM–HM Inequality to $1/x$, $1/y$, $1/z$, we have

$$\frac{1}{x} + \frac{1}{y} + \frac{1}{z} \geq \frac{9}{x + y + z} = \frac{3}{2}.$$

By applying the GM–HM Inequality to the same set of variables, we get

$$\frac{1}{xyz} \geq \left(\frac{3}{x + y + z}\right)^3 = \frac{1}{8}.$$

It follows that

$$\frac{1}{x} + \frac{1}{y} + \frac{1}{z} \geq \frac{3}{2} = 2 - 4\left(\frac{1}{8}\right) \geq 2 - \frac{4}{xyz},$$

with equality if and only if $x = y = z = 2$. Comparing the above inequality with the second equation in the system, we see that $(2, 2, 2)$ is the only solution to the system.

Determine all whole numbers m for which all solutions of the equation $3x^3 - 3x^2 + m = 0$ are rational numbers.

Solution

If $m = 0$, the given equation reduces to $x^3 - x^2 = 0$. The solutions of this equation are 0 (repeated) and 1, both of which are rational. Conversely, we will show that if all solutions of the equation are rational, then $m = 0$.

Suppose that all solutions are rational. Then there is a rational solution of the form $x = a/b$, where a and b are integers, $b > 0$, and a and b are relatively prime. Substituting such a solution into the equation, we get

$$3a^3 - 3a^2b + b^3m = 0.$$

Since b divides the last two terms, it follows that $b \mid 3a^3$. Then, since a and b are relatively prime, we must have $b = 1$ or $b = 3$. If $b = 3$, then $3a^3 - 9a^2 + 27m = 0$, or $a^3 - 3a^2 + 9m = 0$; hence, $3 \mid a$. This is a contradiction, since a and b are relatively prime. Therefore, $b = 1$. We conclude that $m = 3a^2(1 - a)$.

Setting $m = 3a^2(1 - a)$ in the given equation, the equation becomes $x^3 - x^2 - a^3 + a^2 = 0$, or

$$(x - a)(x^2 + ax + a^2 - x - a) = 0.$$

The discriminant of the quadratic factor is $(1 - a)(3a + 1)$. Recalling that a is an integer, we see that the discriminant is non-negative only when $a = 0$ or $a = 1$. In both of these cases, we have $m = 0$.

Determine all positive integers n such that $n! > \sqrt{n^n}$.

Solution

Clearly, $n! = \sqrt{n^n}$ for $n = 1$ and $n = 2$. We show by induction that $n! > \sqrt{n^n}$, or equivalently, $(n!)^2 > n^n$, for all $n \geq 3$. This is clearly true for $n = 3$. Suppose $(n!)^2 > n^n$ for some $n \geq 3$. Then

$$\left((n+1)!\right)^2 = (n+1)^2(n!)^2 > (n+1)^2 n^n. \tag{1}$$

Since $(\frac{n+1}{n})^n = (1 + \frac{1}{n})^n < e \leq 3 < n + 1$, we have $n^n > (n+1)^{n-1}$. Hence,

$$(n+1)^2 n^n > (n+1)^{n+1}. \tag{2}$$

From (1) and (2), it follows that $\left((n+1)!\right)^2 > (n+1)^{n+1}$, completing the induction.

Let $a_1, a_2, \ldots, a_n$ be real numbers such that

$$a_1 + a_2 + \cdots + a_n \geq n^2 \quad \text{and} \quad a_1^2 + a_2^2 + \cdots + a_n^2 \leq n^3 + 1 \,.$$

Prove that $n - 1 \leq a_k \leq n + 1$ for all k.

Solution

First, note that the inequality $a_1 + a_2 + \cdots + a_n \geq n^2$ can be rewritten as

$$(a_1 - n) + (a_2 - n) + \cdots + (a_n - n) \geq 0 \,. \tag{1}$$

Now,

$$(a_1 - n)^2 + (a_2 - n)^2 + \cdots + (a_n - n)^2$$
$$= a_1^2 + a_2^2 + \cdots + a_n^2 - n^3 - 2n((a_1 - n) + (a_2 - n) + \cdots + (a_n - n))$$
$$\leq 1 \,,$$

using (1) and the given inequality $a_1^2 + a_2^2 + \cdots + a_n^2 \leq n^3 + 1$. Thus, we certainly have $(a_k - n)^2 \leq 1$; that is, $n - 1 \leq a_k \leq n + 1$ for all k.

Let a, b, c and α, β, γ be positive real numbers such that $\alpha + \beta + \gamma = 1$. Prove the inequality

$$\alpha a + \beta b + \gamma c + 2\sqrt{(\alpha\beta + \beta\gamma + \gamma\alpha)(ab + bc + ca)} \leq a + b + c.$$

Solution

Since the inequality is homogeneous in a, b, and c, we may assume that $a + b + c = 1$ (without loss of generality). Then, using the AM–GM Inequality,

$$2\sqrt{(\alpha\beta + \beta\gamma + \gamma\alpha)(ab + bc + ca)}$$

$$\leq \alpha\beta + \beta\gamma + \gamma\alpha + ab + bc + ca$$

$$= \frac{(\alpha + \beta + \gamma)^2 - \alpha^2 - \beta^2 - \gamma^2}{2} + \frac{(a + b + c)^2 - a^2 - b^2 - c^2}{2}$$

$$= 1 - \frac{a^2 + \alpha^2}{2} - \frac{b^2 + \beta^2}{2} - \frac{c^2 + \gamma^2}{2}$$

$$\leq 1 - \sqrt{a^2\alpha^2} - \sqrt{b^2\beta^2} - \sqrt{c^2\gamma^2}$$

$$= a + b + c - a\alpha - b\beta - c\gamma.$$

The result follows immediately.

Given the equation $x^{2001} = y^x$,

(a) find all solution pairs (x, y) consisting of positive integers with x prime;

(b) find all solution pairs (x, y) consisting of positive integers.

(Recall that $2001 = 3 \cdot 23 \cdot 29$.)

Solution

Let us solve (b) directly.

If $x = 1$, we find the solution $(x, y) = (1, 1)$. Now, we assume that $x \geq 2$; then $y \geq 2$. Clearly, x and y have the same prime divisors.

Let $x = p_1^{\alpha_1} p_2^{\alpha_2} \cdots p_n^{\alpha_n}$ and $y = p_1^{\beta_1} p_2^{\beta_2} \cdots p_n^{\beta_n}$ be the prime decompositions of x and y. Thus, for each i, we have

$$2001\alpha_i = x\beta_i. \tag{1}$$

If $\gcd(p_i, 2001) = 1$, then, from (1), we deduce that p_i divides α_i; indeed, we even have $p_i^{\alpha_i}$ divides α_i. But it is easy to prove by induction that $2^k > k$ for each positive integer k. Hence, $p_i^{\alpha_i} \geq 2^{\alpha_i} > \alpha_i$. Thus, for each i, we have $\gcd(p_i, 2001) = p_i$.

Since $2001 = 3 \times 23 \times 29$, it follows that $x = 3^\alpha \times 23^\beta \times 29^\gamma$ for some non-negative integers α, β, γ. From (1), if $\alpha > 0$, we deduce (as above) that $3^{\alpha-1}$ divides α, which leads to $\alpha = 1$. Similarly, we have β, $\gamma \in \{0, 1\}$. Thus, x divides 2001.

Conversely, if $2001 = kx$ for some positive integer k, then the given equation $x^{2001} = y^x$ is equivalent to $y = x^k$. Thus, the set of solutions (x, y) is the set of all ordered pairs $(x, x^{2001/x})$, where x is any positive divisor of 2001; that is, $x \in \{1, 3, 23, 29, 69, 87, 667, 2001\}$.

Show that the inequality

$$\sum_{i=1}^{n} i x_i \leq \binom{n}{2} + \sum_{i=1}^{n} x_i^i$$

holds for every integer $n \geq 2$ and all real numbers $x_1, x_2, \ldots, x_n \geq 0$.

Solution

From the AM–GM Inequality, for each non-negative real number x and for each positive integer k, we have

$$x^k + k - 1 = x^k + 1 + \cdots + 1 \geq k \sqrt[k]{x^k \times 1 \times \cdots \times 1} = kx.$$

Equality occurs, for $k \geq 2$, if and only if $x = 1$.

Using this for each x_i and summing leads to

$$\sum_{i=1}^{n} i x_i \leq \sum_{i=1}^{n} \left((i-1) + x_i^i \right) = \binom{n}{2} + \sum_{i=1}^{n} x_i^i,$$

as desired.

Equality occurs if and only if $x_2 = \cdots = x_n = 1$ (and $x_1 \geq 0$ is arbitrary).

Find positive integers x, y, z such that $x > z > 1999 \cdot 2000 \cdot 2001 > y$ and $2000x^2 + y^2 = 2001z^2$.

Solution

If (x, y, z) is such a triple of positive integers, then

$$\frac{2000(x - z)}{z - y} = \frac{z + y}{x + z}.$$

Denoting this rational by $\frac{m}{n}$ (with m, $n \in \mathbb{N}$), we have

$$n(y + z) = m(x + z) \quad \text{and} \quad 2000n(x - z) = m(z - y).$$

These equations are satisfied by the integers

$$x = d(2000n^2 + 2mn - m^2), \quad y = d(m^2 + 4000mn - 2000n^2),$$

$$z = d(m^2 + 2000n^2)$$

for any integer d. Taking $m = 1$, $n = 2$, and $d = 1999 \times 2000$, we find that

$$x = 1999 \times 2000 \times 8003, \quad y = 1999 \times 2000, \quad z = 1999 \times 2000 \times 8001.$$

These integers x, y, and z satisfy all the given conditions.

The equation $x^2 + y^2 + z^2 + u^2 = xyzu + 6$ is given. Find:

(a) at least one solution in positive integers;

(b) at least 33 such solutions;

(c) at least 100 such solutions.

Solution

For all $n \in \mathbb{N}$, $(x, y, z, u) = (1, 2, n, n+1)$ satisfies the given equation since $x^2 + y^2 + z^2 + u^2 = xyzu + 6 = 2n^2 + 2n + 6$. Hence, there are infinitely many solutions in positive integers.

Let x, y, and z denote positive real numbers, each less than 4. Prove that at least one of the numbers $\frac{1}{x} + \frac{1}{4-y}$, $\frac{1}{y} + \frac{1}{4-z}$, and $\frac{1}{z} + \frac{1}{4-x}$ is greater than or equal to 1.

Solution

From the HM–AM Inequality, we have

$$\frac{1}{2}\left(\frac{1}{x} + \frac{1}{4-x}\right) \geq \frac{2}{x + 4 - x} = \frac{1}{2};$$

whence, $\frac{1}{x} + \frac{1}{4-x} \geq 1$. Now, if the three numbers

$$\frac{1}{x} + \frac{1}{4-y}, \quad \frac{1}{y} + \frac{1}{4-z}, \quad \text{and} \quad \frac{1}{z} + \frac{1}{4-x}$$

were all less than 1, their sum S would be less than 3. However,

$$S = \left(\frac{1}{x} + \frac{1}{4-x}\right) + \left(\frac{1}{y} + \frac{1}{4-y}\right) + \left(\frac{1}{z} + \frac{1}{4-z}\right) \geq 3.$$

This contradiction proves the requested result.

Find the integer solutions of $5x^2 - 14y^2 = 11z^2$.

Solution

Clearly, $x = y = z = 0$ is a solution. We show that this is the only solution.

If $x = 0$, then $y = z = 0$. Next note that x and z must have the same parity. Let us now work modulo 8. If x and z are both odd, then $x^2 \equiv z^2 \equiv 1$, which implies that $5x^2 - 11z^2 \equiv 2$. Since $14y^2 \equiv 14 \equiv 6$ if y is odd and $14y^2 \equiv 0$ if y is even, we have a contradiction. Thus, x and z are both even.

Suppose there are solutions in which $x \neq 0$. Let x_0 denote the least positive integer for which there exist $y, z \in \mathbb{Z}$ such that (x_0, y, z) is a solution. Setting $x_0 = 2x_1$, $z = 2z_1$, we get $20x_1^2 - 14y^2 = 44z_1^2$, or $10x_1^2 - 7y^2 = 22z_1^2$, which implies that y is even. Setting $y = 2y_1$, we then have $10x_1^2 - 28y_1^2 = 22z_1^2$, or $5x_1^2 - 14y_1^2 = 11z_1^2$, showing that (x_1, y_1, z_1) is also a solution. This is a contradiction, since $0 < x_1 < x_0$.

Three real numbers a, b, $c \geq 0$ satisfy

$$a^2 \leq b^2 + c^2, \qquad b^2 \leq c^2 + a^2, \qquad c^2 \leq a^2 + b^2 \,.$$

Prove the inequality

$$(a + b + c)(a^2 + b^2 + c^2)(a^3 + b^3 + c^3) \geq 4(a^6 + b^6 + c^6) \,.$$

When does equality hold?

Solution

By the Cauchy-Schwarz Inequality, we have

$$(a + b + c)(a^3 + b^3 + c^3) \geq (a^2 + b^2 + c^2)^2 \,.$$

Thus,

$$(a + b + c)(a^2 + b^2 + c^2)(a^3 + b^3 + c^3) \geq (a^2 + b^2 + c^2)^3 \,.$$

Therefore, it will suffice to prove that

$$(a^2 + b^2 + c^2)^3 \geq 4(a^6 + b^6 + c^6) \,.$$

We have

$$
\begin{aligned}
(a^2 + b^2 + c^2)^3 &= (a^6 + b^6 + c^6) + 6(abc)^2 \\
&\quad + 3(a^2 b^4 + b^2 c^4 + c^2 a^4 + a^4 b^2 + b^4 c^2 + c^4 a^2) \\
&= 4(a^6 + b^6 + c^6) + 12(abc)^2 \\
&\quad + 3(a^2 + b^2 - c^2)(b^2 + c^2 - a^2)(c^2 + a^2 - b^2) \\
&\geq 4(a^6 + b^6 + c^6) \,,
\end{aligned}
$$

since $a^2 \leq b^2 + c^2$, $b^2 \leq c^2 + a^2$, and $c^2 \leq a^2 + b^2$. Equality holds if and only if one of a, b, c is 0 and the other two are equal.

(a) Prove the inequality

$$\frac{(1+x_1)(1+x_2)\cdots(1+x_n)}{1+x_1x_2\cdots x_n} \leq 2^{n-1}, \quad \forall x_1, x_2, \ldots, x_n \in [1, +\infty).$$

(b) When does the equality hold?

Solution

For each $A \subseteq \{1, 2, \ldots, n\}$, let $p(A) = \prod_{i \in A} x_i$. Then

$$(1+x_1)(1+x_2)\ldots(1+x_n) = \sum_A p(A).$$

Note that for all $a, b \geq 1$, we have $a + b \leq 1 + ab$, with equality if and only if $a = 1$ or $b = 1$. It follows that for all $A \subseteq \{1, 2, \ldots, n\}$, we have

$$p(A) + p(\overline{A}) = \prod_{i \in A} x_i + \prod_{i \notin A} x_i \leq 1 + \prod_{i=1}^{n} x_i.$$

Now we obtain the desired inequality as follows:

$$(1+x_1)(1+x_2)\cdots(1+x_n) = \sum_A p(A) = \frac{1}{2}\sum_A \left[p(A) + p(\overline{A})\right]$$

$$\leq \frac{1}{2}\left(1 + \prod_{i=1}^{n} x_i\right)\sum_A 1 = (1 + x_1 x_2 \cdots x_n)2^{n-1}.$$

Equality holds if and only if, for all $A \subseteq \{1, 2, \ldots, n\}$, we have $p(A) = 1$ or $p(\overline{A}) = 1$. Therefore, equality holds if and only if at most one of the x_i is greater than 1.

Prove that, if $0 < a < b < \dfrac{\pi}{2}$, then

(a) $\dfrac{a}{b} < \dfrac{\sin a}{\sin b}$;

(b) $\dfrac{\sin a}{\sin b} < \dfrac{\pi}{2}\dfrac{a}{b}$.

Solution

(a) Let $I = \left(0, \dfrac{\pi}{2}\right)$ and $f(x) = \dfrac{\sin x}{x}$. Then $f'(x) = \dfrac{\cos x}{x^2}(x - \tan x)$ for all $x \in I$. It is well known that $x < \tan x$ for all $x \in I$, so that $f'(x) < 0$, which means that f is decreasing on I. Therefore, if $0 < a < b < \dfrac{\pi}{2}$, then $f(a) > f(b)$, which is equivalent to $\dfrac{a}{b} < \dfrac{\sin a}{\sin b}$.

(b) It is well known that $\dfrac{2}{\pi}x < \sin x < x$ if $x \in \left(0, \dfrac{\pi}{2}\right)$. Therefore, if $0 < a < b < \dfrac{\pi}{2}$, then $\sin a < a$ and $\dfrac{2}{\pi}b < \sin b$. Thus, $\dfrac{\sin a}{\sin b} < \dfrac{\pi}{2}\dfrac{a}{b}$.

Determine all positive integers m and n such that

$$m^2 - n^2 = 270.$$

Solution

We show that, more generally, the equation $m^2 - n^2 = 2k$ has no integer solutions if k is odd.

Suppose m and n are positive integers such that $m^2 - n^2 = 2k$, where k is an odd integer. Clearly, m and n have the same parity. Hence, $m + n$ and $m - n$ have the same parity. Since $(m + n)(m - n) = m^2 - n^2 = 2k$ is even, both $m + n$ and $m - n$ must be even, which is impossible, since $4 \nmid 2k$.

The positive integers a, b, and c satisfy $\frac{1}{a} + \frac{1}{b} + \frac{1}{c} < 1$. Prove that

$$\frac{1}{a} + \frac{1}{b} + \frac{1}{c} \leq \frac{41}{42}.$$

Solution

Let $S = \frac{1}{a} + \frac{1}{b} + \frac{1}{c}$. Due to complete symmetry, we may assume that $a \leq b \leq c$. Clearly, $a \geq 2$.

If $a \geq 4$, then $S \leq \frac{1}{4} + \frac{1}{4} + \frac{1}{4} = \frac{3}{4} < \frac{41}{42}$. If $a = 3$, then $\frac{1}{b} + \frac{1}{c} < \frac{2}{3}$. Since $b = c = 3$ clearly yields a contradiction, we must have $b \geq 3$ and $c \geq 4$. Then $S \leq \frac{1}{3} + \frac{1}{3} + \frac{1}{4} = \frac{11}{12} < \frac{41}{42}$.

It remains to consider the case when $a = 2$. Since $\frac{1}{b} + \frac{1}{c} < \frac{1}{2}$, we have $b \geq 3$. If $b = 3$, then $\frac{1}{c} < \frac{1}{6}$ yields $c \geq 7$, and hence, $S \leq \frac{1}{2} + \frac{1}{3} + \frac{1}{7} = \frac{41}{42}$. Now suppose that $b \geq 4$. Since $b = c = 4$ clearly yields a contradiction, we must have $b \geq 4$ and $c \geq 5$. Then $S \leq \frac{1}{2} + \frac{1}{4} + \frac{1}{5} = \frac{19}{20} < \frac{41}{42}$.

We see that in all cases, $S \leq \frac{41}{42}$, and that equality holds if and only if $(a, b, c) = (2, 3, 7)$.

Prove that, if $m \cdot s = 2000^{2001}$ where $m, s \in \mathbb{Z}$, then the equation $mx^2 - sy^2 = 3$ has no solution in $\mathbb{Z}$.

Solution

Let $m, s \in \mathbb{Z}$ such that $m \cdot s = 2001^{2001}$. Suppose, for the purpose of contradiction, that $mx^2 - sy^2 = 3$ for some integers x and y.

Note that $ms = 2^{8004}5^{6003}$. Since ms is even, m and s cannot both be odd. If s is even, then m must be odd, since $mx^2 = 3 + sy^2$ is odd. Thus, $m = 5^{\alpha}$ for some integer α with $0 \leq \alpha \leq 6003$, and $s = 2^{8004}5^{6003-\alpha}$. It follows that $5^{\alpha}x^2 = 3 + 2^{8004}6^{6003-\alpha}y^2$. Modulo 4, this yields $x^2 \equiv 3$, which is a contradiction, since a square is congruent to 0 or 1 modulo 4.

If m is even, then s is odd. Hence, $s = 5^{\beta}$ for some integer β with $0 \leq \beta \leq 6003$, and $m = 2^{8004}5^{6003-\beta}$. Thus, $3 + 5^{\beta}y^2 = 2^{8004}5^{6003-\beta}x^2$. This implies that β cannot lie strictly between 0 and 6003 (since 3 is not a multiple of 5). If $\beta = 0$, then $y^2 \equiv -3 \equiv 2 \pmod 5$; if $\beta = 6003$, then $x^2 \equiv 3 \pmod 5$ (since $2^{8004} = 4^{4002} \equiv (-1)^{4002} \equiv 1 \pmod 5$). We again have a contradiction, since a square is congruent to 0, 1 or 4, modulo 5.

Let $x \geq 0$, $y \geq 0$ be real numbers with $x + y = 2$. Prove that

$$x^2 y^2 (x^2 + y^2) \leq 2.$$

Solution

By the AM–GM Inequality, we have $\sqrt{xy} \leq \frac{x+y}{2} = \frac{2}{2} = 1$; whence $xy \leq 1$. Then, using the AM–GM Inequality again, we get

$$x^2 y^2 = \sqrt{x^4 y^4} \leq \frac{x^4 y^4 + 1}{2} \leq \frac{x^3 y^3 + 1}{2};$$

whence, $x^2 y^2 (2 - xy) \leq 1$. We also have

$$x^2 + y^2 = (x + y)^2 - 2xy = 4 - 2xy = 2(2 - xy).$$

Now $x^2 y^2 (x^2 + y^2) = 2x^2 y^2 (2 - xy) \leq 2.$

For each positive integer n, determine, with proof, all positive integers m such that there exist positive integers $x_1 < x_2 < \cdots < x_n$ which satisfy
$$\frac{1}{x_1} + \frac{2}{x_2} + \frac{3}{x_3} + \cdots + \frac{n}{x_n} = m.$$

Solution

Let n be any fixed positive integer. We first suppose that there exist positive integers $x_1 < x_2 < \cdots < x_n$ such that $\sum\limits_{i=1}^{n} \frac{i}{x_i}$ is a positive integer m. Since $x_1, x_2, \ldots, x_n$ are positive integers and $x_1 < x_2 < \cdots < x_n$, we must have $x_i \geq i$ for all $i = 1, 2, \ldots, n$. Then

$$m = \sum_{i=1}^{n} \frac{i}{x_i} \leq \sum_{i=1}^{n} 1 = n.$$

Next we shall show that, for all integers m such that $1 \leq m \leq n$, such numbers x_i do exist. Indeed, for $m = n$, set $x_i = i$ and for $m = 1$, set $x_i = in$. It can be verified that $\sum\limits_{i=1}^{n} \frac{i}{x_i} = m$ in both cases. For $1 < m < n$, we write

$$\sum_{i=1}^{n} \frac{i}{x_i} = \underbrace{\frac{1}{x_1} + \frac{2}{x_2} + \cdots + \frac{m-1}{x_{m-1}}}_{m-1 \text{ terms}} + \underbrace{\frac{m}{x_m} + \cdots + \frac{n}{x_n}}_{n-m+1 \text{ terms}}$$

and note that, in order to get $\sum\limits_{i=1}^{n} \frac{i}{x_i} = m$, it suffices to make the first sum equal to $m - 1$ by setting $x_i = i$ for $1 \leq i \leq m - 1$, and the second sum equal to 1 by setting $x_i = i(n - m + 1)$ for $m \leq i \leq n$. It is easy to see that $x_1 < x_2 < \cdots < x_n$ in all cases, and the proof is complete.

Let $a_1 = 1$, $a_{n+1} = \dfrac{a_n}{n} + \dfrac{n}{a_n}$ for $n = 1, 2, 3, \ldots$. Find the greatest integer less than or equal to a_{2000}. Be sure to prove your claim.

Solution

Direct computation gives $a_2 = a_3 = 2$. Therefore, $\sqrt{3} < a_3 < 3/\sqrt{2}$. We prove by induction that $\sqrt{n} < a_n < n/\sqrt{n-1}$ for each integer $n \geq 3$. Assume that this holds for some $n \geq 3$. Then we have

$$0 < \sqrt{n} < a_n < \frac{n}{\sqrt{n-1}} < n.$$

Note that $a_{n+1} = f(a_n)$, where $f(x) = \dfrac{x}{n} + \dfrac{n}{x}$. The function f is decreasing on $(0, n)$, since $f'(x) = \dfrac{x^2 - n^2}{nx^2} < 0$ for $0 < x < n$. Therefore,

$$f\left(\frac{n}{\sqrt{n-1}}\right) < f(a_n) < f(\sqrt{n}).$$

But $f(\sqrt{n}) = \dfrac{1}{\sqrt{n}} + \sqrt{n} = \dfrac{n+1}{\sqrt{n}}$ and

$$f\left(\frac{n}{\sqrt{n-1}}\right) = \frac{1}{\sqrt{n-1}} + \sqrt{n-1} = \frac{n}{\sqrt{n-1}} > \sqrt{n+1}.$$

Thus, $\sqrt{n+1} < a_{n+1} < (n+1)/\sqrt{n}$, which ends the induction.

Now $44 < \sqrt{2000} < a_{2000} < \dfrac{2000}{\sqrt{1999}} < 45$. Therefore, $\lfloor a_{2000} \rfloor = 44$.

Let x and y be positive real numbers such that $xy = 1$. Prove that

$$\frac{x}{y} + \frac{y}{x} \geq 2.$$

Solution

The condition $xy = 1$ is unnecessary. For any positive real numbers x and y,

$$\frac{x}{y} + \frac{y}{x} = \left(\sqrt{\frac{x}{y}} - \sqrt{\frac{y}{x}}\right)^2 + 2 \geq 2.$$

Let x, y, a be real numbers such that

$$x + y = x^3 + y^3 = x^5 + y^5 = a.$$

Determine all the possible values of a.

Solution

The possible values of a are 0, 1, -1, 2, and -2.

Observe that $a = 0$ when $y = -x$, $a = 1$ when $x = 0$ and $y = 1$, $a = -1$ when $x = 0$ and $y = -1$, $a = 2$ when $x = y = 1$, and $a = -2$ when $x = y = -1$.

Conversely, suppose that $x + y = x^3 + y^3 = x^5 + y^5 = a \neq 0$ for some real numbers x and y. We show that a must be 1, -1, 2, or -2.

Let $p = xy$. Then $a = x^3 + y^3 = (x + y)^3 - 3xy(x + y) = a^3 - 3ap$, which yields $a^2 = 1 + 3p$. Also,

$$a = x^5 + y^5 = (x + y)^5 - 5xy(x^3 + y^3) - 10xy(x + y) = a^5 - 15ap,$$

and hence, $a^4 = 1 + 15p$. Thus, $1 + 15p = (1 + 3p)^2$, which is easily solved to get $p = 0$ or $p = 1$.

If $p = 0$, then the equation $a^2 = 1 + 3p$ gives $a = 1$ or $a = -1$. If $p = 1$, then the same equation gives $a = 2$ or $a = -2$.

Prove that the following inequality holds for every triple (a, b, c) of non-negative real numbers with $a^2 + b^2 + c^2 = 1$:

$$\frac{a}{b^2 + 1} + \frac{b}{c^2 + 1} + \frac{c}{a^2 + 1} \geq \frac{3}{4}\left(a\sqrt{a} + b\sqrt{b} + c\sqrt{c}\right)^2 .$$

When does equality hold?

First we note that if x_1, x_2, x_3 are any real numbers and w_1, w_2, w_3 are any positive real numbers, then, by the Cauchy-Schwarz Inequality,

$$\left(\sum_{i=1}^{3} x_i\right)^2 = \left(\sum_{i=1}^{3} \sqrt{w_i}\,\frac{x_i}{\sqrt{w_i}}\right)^2 \leq \left(\sum_{i=1}^{3} w_i\right)\left(\sum_{i=1}^{3} \frac{x_i^2}{w_i}\right). \qquad (1)$$

Now we let $x_1 = a\sqrt{a}$, $x_2 = b\sqrt{b}$, $x_3 = c\sqrt{c}$, $w_1 = a^2(b^2 + 1)$, $w_2 = b^2(c^2 + 1)$, and $w_3 = c^2(a^2 + 1)$. Then

$$\sum_{i=1}^{3} \frac{x_i^2}{w_i} = \frac{\left(a\sqrt{a}\right)^2}{a^2(b^2 + 1)} + \frac{\left(b\sqrt{b}\right)^2}{b^2(c^2 + 1)} + \frac{\left(c\sqrt{c}\right)^2}{c^2(a^2 + 1)}$$

$$= \frac{a}{b^2 + 1} + \frac{b}{c^2 + 1} + \frac{c}{a^2 + 1}, \qquad (2)$$

and

$$\sum_{i=1}^{3} w_i = a^2 b^2 + b^2 c^2 + c^2 a^2 + a^2 + b^2 + c^2 = a^2 b^2 + b^2 c^2 + c^2 a^2 + 1 .$$

Since

$$a^2 b^2 + b^2 c^2 + c^2 a^2 \leq a^4 + b^4 + c^4 = (a^2 + b^2 + c^2)^2 - 2(a^2 b^2 + b^2 c^2 + c^2 a^2),$$

we have

$$a^2 b^2 + b^2 c^2 + c^2 a^2 \leq \tfrac{1}{3}(a^2 + b^2 + c^2)^2 = \tfrac{1}{3} .$$

Thus, $\sum_{i=1}^{3} w_i \leq \frac{1}{3} + 1 = \frac{4}{3}$. Using this result along with (2) in (1), we get

$$\left(a\sqrt{a} + b\sqrt{b} + c\sqrt{c}\right)^2 \leq \frac{4}{3}\left(\frac{a}{b^2 + 1} + \frac{b}{c^2 + 1} + \frac{c}{a^2 + 1}\right),$$

which immediately yields the desired result.

We have equality if and only if $a = b = c = \frac{1}{3}\sqrt{3}$.

Find all positive integers z for which the equation

$$x(x + z) = y^2$$

has no solutions x, y that are positive integers.

Solution

We will say that a positive integer z is *bad* if the equation

$$x(x + z) = y^2 \tag{1}$$

has no solution in positive integers (and z is *good* otherwise). We will prove that the bad integers, for which the problem is asking, are 1, 2 and 4.

First note that if (x, y, z) is a solution of (1) in positive integers, then $(2x, 2y, 2z)$ is also a solution. Therefore, if z is good, then $2z$ is also good.

If $z = 2n + 1$, with $n \geq 1$, then z is good because it suffices to choose $x = n^2$ and $y = n(n + 1)$ to obtain a solution of (1) in positive integers. It follows that each positive integer which is not a power of 2 is good.

For each positive integer x, we have $x^2 < x(x + 4) < (x + 2)^2$. If $x(x + 4) = y^2$, then we must have $x(x + 4) = (x + 1)^2$. But this equation is equivalent to $2x = 1$, which has no integer solutions. Thus, $z = 4$ is bad, from which we deduce that 1, 2, and 4 are bad.

Clearly, $(1, 3, 8)$ is a solution of (1), which means that $z = 8$ is good. Then $z = 2^n$ is good for all integers $n \geq 3$, and the proof is complete.

Let x and y be any two real numbers. Prove that

$$3(x + y + 1)^2 + 1 \geq 3xy.$$

Under what conditions does equality hold?

Solution

For any real numbers X and Y,

$$X^2 + Y^2 + XY = \left(X + \tfrac{1}{2}Y\right)^2 + \tfrac{3}{4}Y^2 \geq 0,$$

and equality holds if and only if $X = Y = 0$.

Let x and y be any two real numbers. Letting $X = x + \tfrac{2}{3}$ and $Y = y + \tfrac{2}{3}$ above, we obtain

$$\left(x + \tfrac{2}{3}\right)^2 + \left(y + \tfrac{2}{3}\right)^2 + \left(x + \tfrac{2}{3}\right)\left(y + \tfrac{2}{3}\right) \geq 0.$$

Expanding and multiplying by 3 gives

$$3x^2 + 3y^2 + 3xy + 6x + 6y + 4 \geq 0.$$

This may be written as

$$3(x + y + 1)^2 - 3xy + 1 \geq 0,$$

from which we arrive at the desired inequality.

Equality holds if and only if $x + \tfrac{2}{3} = y + \tfrac{2}{3} = 0$; that is, if and only if $x = y = -\tfrac{2}{3}$.

Determine all triples of positive integers a, b, c such that a^2+1 and b^2+1 are prime numbers satisfying $(a^2 + 1)(b^2 + 1) = c^2 + 1$.

Solution

The only solution is $(a, b, c) = (1, 2, 3)$.
Suppose that a, b, c satisfy

$$(a^2 + 1)(b^2 + 1) = c^2 + 1, \qquad (1)$$

where $a^2 + 1 = p$ and $b^2 + 1 = q$ are both prime. If $p = q$, then from (1) we obtain $p^2 = c^2 + 1$, or $(p - c)(p + c) = 1$, which is clearly impossible. Thus, $p \neq q$. Without loss of generality, we may assume $p < q$. Then $a < b < c$.

Rewriting (1) as $a^2 b^2 + a^2 + b^2 = c^2$, we have

$$(c - a)(c + a) = c^2 - a^2 = b^2(a^2 + 1) = b^2 p \qquad (2)$$
$$\text{and} \quad (c - b)(c + b) = c^2 - b^2 = a^2(b^2 + 1) = a^2 q. \qquad (3)$$

Since p and q are primes, we deduce from (2) and (3) that $p \mid (c - a)$ or $p \mid (c + a)$, and $q \mid (c - b)$ or $q \mid (c + b)$. Thus, we have four possible cases:

Case 1. $p \mid (c - a)$ and $q \mid (c - b)$.
Then $pq \mid (c - a)(c - b)$. Hence, $(c^2 + 1) \mid (c - a)(c - b)$, which is impossible, since $0 < (c - a)(c - b) < c^2 < c^2 + 1$.

Case 2. $p \mid (c + a)$ and $q \mid (c - b)$.
Then $pq \mid (c + a)(c - b)$. Hence, $(c^2 + 1) \mid (c + a)(c - b)$, which is impossible, since

$$c^2 + 1 - (c + a)(c - b) = bc - ac + ab + 1 = c(b - a) + ab + 1 > 0.$$

Case 3. $p \mid (c - a)$ and $q \mid (c + b)$.
Then $(c^2 + 1) \mid (c - a)(c + b)$. Hence, $(c - a)(c + b) = k(c^2 + 1)$ for some $k \in \mathbb{N}$. However, $(c - a)(c + b) < c(2c) = 2c^2 < 2(c^2 + 1)$; whence, $k = 1$. Therefore, $(c - a)(c + b) = c^2 + 1$, which can be rewritten as $(b - a)(c - a) = a^2 + 1 = p$.

Since $b - a < c - a$, we must have $b - a = 1$ and $c - a = p$. Since $b = a + 1$, we see that a and b have opposite parity; thus, $p = a^2 + 1$ and $q = b^2 + 1$ must also have opposite parity. Since $p < q$, we conclude that $p = 2$. It follows that $a = 1$ and $b = 2$, and we obtain the solution $(a, b, c) = (1, 2, 3)$.

Case 4. $p \mid (c + a)$ and $q \mid (c + b)$.

Then $(c^2 + 1) \mid (c + a)(c + b)$. Hence, $(c + a)(c + b) = m(c^2 + 1)$ for some $m \in \mathbb{N}$. However,

$$
\begin{aligned}
(c + a)(c + b) \ &< \ (c + a)(c + b) + (c - a)(c - b) + (a - b)^2 + a^2 b^2 \\
&= \ 2c^2 + a^2 + b^2 + a^2 b^2 \\
&= \ 3c^2 < 3(c^2 + 1) \,,
\end{aligned}
$$

where we have used (1) in the second-last step. Thus, we see that $m = 2$. Then, from $(c + a)(c + b) = 2(c^2 + 1)$, we obtain

$$(a + b)c + ab \ = \ c^2 + 2 \,. \tag{4}$$

If $p \neq 2$, then both p and q are odd, which implies that a and b are both even. Hence, c is also even. Let $a = 2a_1$, $b = 2b_1$, and $c = 2c_1$. Then (4) becomes $4(a_1 + b_1)c_1 + 4a_1 b_1 = 4c_1^2 + 2$ or $2(a_1 + b_1)c_1 + 2a_1 b_1 = 2c_1^2 + 1$, which is clearly impossible. Hence, $p = 2$ and $a = 1$. Substituting into (3), we then have $(c - b)(c + b) = q$. Hence, $c - b = 1$ and $c + b = q$. Solving, we have $b = (q - 1)/2 = b^2/2$, from which it follows that $b = 2$, and again we are led to the solution $(a, b, c) = (1, 2, 3)$.

This completes the proof.

Prove that, for every integer $n \geq 3$ and every sequence of positive numbers $x_1, x_2, \ldots, x_n$, at least one of the following two inequalities is satisfied:

$$\sum_{i=1}^{n} \frac{x_i}{x_{i+1} + x_{i+2}} \geq \frac{n}{2}, \qquad \sum_{i=1}^{n} \frac{x_i}{x_{i-1} + x_{i-2}} \geq \frac{n}{2}.$$

(Note: Here $x_{n+1} = x_1$, $x_{n+2} = x_2$, $x_0 = x_n$, and $x_{-1} = x_{n-1}$.)

Solution

Denote the sums in the inequalities above by S_1 and S_2, respectively. If $S_1 < n/2$ and $S_2 < n/2$, then $S_1 + S_2 < n$. Therefore, it will be sufficient to prove that $S_1 + S_2 \geq n$.

We start with the inequality $\alpha + \frac{1}{\alpha} \geq 2$, which is true for all positive real numbers α. For each $i \in \{1, 2, \ldots, n\}$, let $\alpha_i = \frac{x_{i-1} + x_i}{x_i + x_{i+1}}$. Then $\alpha_i + \frac{1}{\alpha_i} \geq 2$ for each i, and hence,

$$2n \leq \sum_{i=1}^{n} \left(\alpha_i + \frac{1}{\alpha_i} \right) = \sum_{i=1}^{n} \left(\alpha_i + \frac{1}{\alpha_{i+1}} \right),$$

where $\sum_{i=1}^{n} \frac{1}{\alpha_i} = \sum_{i=1}^{n} \frac{1}{\alpha_{i+1}}$ because the sum is cyclic. Thus,

$$\begin{aligned}
2n &\leq \sum_{i=1}^{n} \left(\frac{x_{i-1} + x_i}{x_i + x_{i+1}} + \frac{x_{i+1} + x_{i+2}}{x_i + x_{i+1}} \right) \\
&= \sum_{i=1}^{n} \left(\frac{x_i + x_{i+1}}{x_i + x_{i+1}} + \frac{x_{i-1} + x_{i+2}}{x_i + x_{i+1}} \right) = n + \sum_{i=1}^{n} \frac{x_{i-1} + x_{i+2}}{x_i + x_{i+1}}.
\end{aligned}$$

Then

$$\begin{aligned}
n &\leq \sum_{i=1}^{n} \frac{x_{i-1}}{x_i + x_{i+1}} + \sum_{i=1}^{n} \frac{x_{i+2}}{x_i + x_{i+1}} \\
&= \sum_{i=1}^{n} \frac{x_i}{x_{i+1} + x_{i+2}} + \sum_{i=1}^{n} \frac{x_i}{x_{i-2} + x_{i-1}} = S_1 + S_2,
\end{aligned}$$

where we have again made use of the cyclic nature of the sums to shift the index.

Let $a_1, a_2, \ldots, a_n$ and $b_1, b_2, \ldots, b_n$ be real numbers between 1001 and 2002 inclusive. Suppose $\sum_{i=1}^{n} a_i^2 = \sum_{i=1}^{n} b_i^2$. Prove that

$$\sum_{i=1}^{n} \frac{a_i^3}{b_i} \leq \frac{17}{10} \sum_{i=1}^{n} a_i^2 .$$

Determine when equality holds.

Solution

For each i, we have

$$\frac{1}{2} = \frac{1001}{2002} \leq \frac{a_i}{b_i} \leq \frac{2002}{1001} = 2 ,$$

and therefore $(2a_i - b_i)(2b_i - a_i) \geq 0$; that is,

$$5a_i b_i \geq 2(a_i^2 + b_i^2) . \tag{7}$$

Multiplying this inequality by a_i / b_i, we get

$$5a_i^2 \geq 2\frac{a_i^3}{b_i} + 2a_i b_i . \tag{8}$$

From (7), we have $2a_i b_i \geq \frac{4}{5}(a_i^2 + b_i^2)$. Using this inequality in (8), we obtain

$$5a_i^2 \geq 2\frac{a_i^3}{b_i} + \frac{4}{5}(a_i^2 + b_i^2) ,$$

which may be rewritten as

$$\frac{a_i^3}{b_i} \leq \frac{21}{10}a_i^2 - \frac{2}{5}b_i^2 . \tag{9}$$

Note that equality occurs in (9) if and only if $b_i = 2a_i$ or $a_i = 2b_i$; that is, if and only if $(a_i, b_i) = (1001, 2002)$ or $(a_i, b_i) = (2002, 1001)$.

Summing over i in (9) and recalling that $\sum_{i=1}^{n} b_i^2 = \sum_{i=1}^{n} a_i^2$, we get

$$\sum_{i=1}^{n} \frac{a_i^3}{b_i} \leq \frac{21}{10} \sum_{i=1}^{n} a_i^2 - \frac{2}{5} \sum_{i=1}^{n} a_i^2 = \frac{17}{10} \sum_{i=1}^{n} a_i^2 ,$$

as desired.

Equality occurs if and only if, for each i, either $(a_i, b_i) = (1001, 2002)$ or $(a_i, b_i) = (2002, 1001)$. The condition $\sum_{i=1}^{n} a_i^2 = \sum_{i=1}^{n} b_i^2$ can be rewritten as $1001^2 p + (n - p)2002^2 = 2002^2 p + 1001^2 (n - p)$, which is $p = \frac{1}{2}n$. Thus, equality occurs if and only if n is even and $(a_i, b_i) = (1001, 2002)$ for half of the subscripts i while $(a_i, b_i) = (2002, 1001)$ for the other half.

Find all values of n for which all solutions of the equation $x^3 - 3x + n = 0$ are integers.

Solution

If the given equation has an integer root α, then $n = -\alpha^3 + 3\alpha$ is an integer too. Therefore, the values of n that we seek must all be integers.

Let $f(x) = x^3 - 3x$. The given equation is then $f(x) = -n$. Straight-forward computations show that f is increasing on $(-\infty, -1]$ and $[1, +\infty)$, and decreasing on $[-1, 1]$. Moreover, $f(-1) = 2$ and $f(1) = -2$. Thus, the equation $f(x) = -n$ has three real roots if and only if $|n| \leq 2$. Therefore, $n \in \{-2, -1, 0, 1, 2\}$. Furthermore, one of the integer solutions has to be -1, 0, or 1; thus, $n \in \{f(-1), f(0), f(1)\} = \{-2, 0, 2\}$. Direct checking shows that the desired values are $n = -2$ and $n = 2$.

Find all positive integers m, n, where n is odd, that satisfy

$$\frac{1}{m} + \frac{4}{n} = \frac{1}{12}.$$

Solution

In the following, all letters stand for positive integers.
Suppose m and n satisfy the given conditions; that is, n is odd and

$$12n + 48m = mn. \tag{1}$$

Then $4 \mid mn$ and, since $(4, n) = 1$, we have $4 \mid m$; that is, $m = 4t$. Then (1) reduces to

$$3n + 48t = tn. \tag{2}$$

From (2), we have $n \mid 48t$. Since $48t = 16 \cdot 3t$ and $(n, 16) = 1$, we have $n \mid 3t$; that is, $3t = nu$. Then (2) reduces to

$$9 + 48u = nu. \tag{3}$$

Therefore, $u \mid 9$; that is, $u = 1$, $u = 3$, or $u = 9$. Then (3) yields $n = 57$, $n = 51$, or $n = 49$, respectively. From (1), we get $m = 76$, $m = 204$, or $m = 588$, respectively.
◀Thus, we find that the only possible pairs (m, n) are $(76, 57)$, $(204, 51)$, and $(588, 49)$. All three satisfy the required conditions.

Find all positive real solutions to the equation

$$x + \left\lfloor \frac{x}{6} \right\rfloor = \left\lfloor \frac{x}{2} \right\rfloor + \left\lfloor \frac{2x}{3} \right\rfloor,$$

where $\lfloor t \rfloor$ denotes the largest integer less than or equal to the real number t.

Solution

Clearly x must be an integer. Let $x = 6k+r$, where k and r are integers and $0 \le r \le 5$. Then the given equation becomes

$$7k + r = 7k + \left\lfloor \frac{r}{2} \right\rfloor + \left\lfloor \frac{2r}{3} \right\rfloor, \quad \text{or} \quad r = \left\lfloor \frac{r}{2} \right\rfloor + \left\lfloor \frac{2r}{3} \right\rfloor.$$

We can easily check that this equation is satisfied for $r \in \{0, 2, 3, 4, 5\}$, but not for $r = 1$. Thus, the set of all real solutions is $\{x \in \mathbb{Z} \mid x \not\equiv 1 \pmod 6\}$, and the set of all positive real solutions is $\{x \in \mathbb{N} \mid x \not\equiv 1 \pmod 6\}$.

Let a, b, and c be positive numbers, and let n and k be positive integers. Prove the inequality:

$$\frac{a^{n+k}}{b^n} + \frac{b^{n+k}}{c^n} + \frac{c^{n+k}}{a^n} \geq a^k + b^k + c^k \,.$$

Solution

The proposed inequality is equivalent to

$$c^n a^{2n+k} + a^n b^{2n+k} + b^n c^{2n+k} \geq b^n c^n a^{n+k} + c^n a^n b^{n+k} + a^n b^n c^{n+k} \,,$$

or

$$c^n a^{n+k}(a^n - b^n) + a^n b^{n+k}(b^n - c^n) + b^n c^{n+k}(c^n - a^n) \geq 0 \,.$$

Since the left side L remains the same after a circular permutation of a, b, c, we may consider only the cases $a \geq b \geq c$ and $a \geq c \geq b$.

In the former case, we use $c^n - a^n = (c^n - b^n) + (b^n - a^n)$ to get

$$L = c^n(a^n - b^n)(a^{n+k} - b^n c^k) + b^n(b^n - c^n)(a^n b^k - c^{n+k}) \,.$$

The first term on the right side is non-negative because $a^n \geq b^n$ and $a^k \geq c^k$; the second term is non-negative because $b^n \geq c^n$, $a^n \geq c^n$, and $b^k \geq c^k$. Thus, $L \geq 0$.

In the latter case, similarly, we use $a^n - b^n = (a^n - c^n) + (c^n - b^n)$ to get

$$L = c^n(a^n - c^n)(a^{n+k} - b^n c^k) + a^n(c^n - b^n)(c^n a^k - b^{n+k}) \,,$$

which is again the sum of two non-negative terms. Thus, $L \geq 0$ in all cases, and the result follows.

What is the maximal value of the expression $a + b + c + abc$, if a, b, c are non-negative numbers such that $a^2 + b^2 + c^2 + abc \leq 4$?

Solution

Let $f(a, b, c) = a + b + c + abc$ and $g(a, b, c) = a^2 + b^2 + c^2 + abc$. Let S be the set of the triples (a, b, c) of non-negative numbers such that $g(a, b, c) \leq 4$.

First, it is clear that $S \subset [0, 2]^3$ and that S is a closed set. Since f is continuous, the desired maximum exists. Let M denote this maximum. Since $f(1, 1, 1) = 4$, we must have

$$M \geq 4. \qquad (1)$$

Let $(x, y, z) \in S$ such that $f(x, y, z) = M$. If one of x, y, z is 2, say $x = 2$, then $y = z = 0$ and $f(x, y, z) = 2$, which is a contradiction since $f(x, y, z) = M \geq 4$. If one of x, y, z is 0, say $x = 0$, then $y^2 + z^2 \leq 4$ and, using the inequality between the arithmetic and quadratic means, we have

$$f(x, y, z) = y + z \leq 2\sqrt{\frac{y^2 + z^2}{2}} \leq 2\sqrt{2} < M,$$

a contradiction. Thus,

$$0 < x, y, z < 2. \qquad (2)$$

Now we will prove that, if $x \neq y$, then $\left(\frac{1}{2}(x + y), \frac{1}{2}(x + y), z\right) \in S$ and $f\left(\frac{1}{2}(x + y), \frac{1}{2}(x + y), z\right) > f(x, y, z)$.

We have

$$g\left(\tfrac{1}{2}(x + y), \tfrac{1}{2}(x + y), z\right)$$
$$= \left(\tfrac{1}{2}(x + y)\right)^2 + \left(\tfrac{1}{2}(x + y)\right)^2 + z^2 + \left(\tfrac{1}{2}(x + y)\right)^2 z$$
$$= x^2 + y^2 + z^2 + xyz - \tfrac{1}{4}(x - y)^2(2 - z)$$
$$= g(x, y, z) - \tfrac{1}{4}(x - y)^2(2 - z) \leq g(x, y, z),$$

since $z < 2$. This proves that $\left(\frac{1}{2}(x + y), \frac{1}{2}(x + y), z\right) \in S$. Moreover,

$$f\left(\tfrac{1}{2}(x + y), \tfrac{1}{2}(x + y), z\right) - f(x, y, z) = \left[\left(\tfrac{1}{2}(x + y)\right)^2 - xy\right]z > 0,$$

by the AM–GM Inequality and (2). Thus,

$$f\left(\tfrac{1}{2}(x + y), \tfrac{1}{2}(x + y), z\right) > f(x, y, z),$$

as claimed.

Since $f(x, y, z) = M$ is the maximal value of f, this proves that $x = y$. We can prove in the same way that $x = z$, so that $x = y = z$. Let t be this common value. Then $3t^2 + t^3 \leq 4$, which forces $t \leq 1$. On the other hand, for $t \leq 1$, we clearly have $f(t, t, t) = 3t + t^3 \leq 4$. Then $M \leq 4$.

In view of (1), it follows that $M = 4$.

Let x, y, and z be real numbers such that

$$x + y + z = 3 \quad \text{and} \quad xy + yz + xz = a$$

(a is a real parameter). Determine the value of the parameter a for which the difference between the maximum and minimum possible values of x equals 8.

Solution

We have $y + z = 3 - x$ and $yz = a - x(3 - x)$. It is easy to verify that the system

$$y + z = s,$$
$$yz = p,$$

with unknowns y and z, has a real solution if and only if $s^2 - 4p \geq 0$. Hence, the unique condition we have to satisfy is $(3 - x)^2 \geq 4(a - 3x + x^2)$, or $3(x - 1)^2 \leq 4(3 - a)$. That is, $a \leq 3$ and

$$1 - 2\sqrt{1 - \tfrac{1}{3}a} \leq x \leq 1 + 2\sqrt{1 - \tfrac{1}{3}a}.$$

The difference between the maximum and minimum possible values of x equals 8 if and only if $4\sqrt{1 - \tfrac{1}{3}a} = 8$, which implies that $a = -9$.

Let a, b, and c be real numbers such that $a^2 + b^2 + c^2 = 1$. Prove the inequality

$$\frac{a^2}{1 + 2bc} + \frac{b^2}{1 + 2ca} + \frac{c^2}{1 + 2ab} \geq \frac{3}{5}.$$

Solution

In view of the inequality $2xy \leq x^2 + y^2$ and the observation that $1 + 2bc = a^2 + (b + c)^2 > 0$, etc., we see that

$$\frac{a^2}{1 + 2bc} + \frac{b^2}{1 + 2ca} + \frac{c^2}{1 + 2ab} \geq \frac{a^2}{1 + b^2 + c^2} + \frac{b^2}{1 + c^2 + a^2} + \frac{c^2}{1 + a^2 + b^2}$$

$$= \frac{a^2}{2 - a^2} + \frac{b^2}{2 - b^2} + \frac{c^2}{2 - c^2}.$$

Let $f : (0, 1) \to \mathbb{R}$ be defined as $f(x) = x/(2 - x)$. Then

$$f'(x) = \frac{2}{(2 - x)^2} \quad \text{and} \quad f''(x) = \frac{4}{(2 - x)^3}.$$

Thus, f is convex, since $f''(x) > 0$. Using Jensen's Inequality, we get

$$f\left(\frac{x + y + z}{3}\right) \leq \frac{f(x) + f(y) + f(z)}{3}.$$

Taking $x = a^2$, $y = b^2$, and $z = c^2$ gives $f(a^2) + f(b^2) + f(c^2) \geq 3f(\frac{1}{3})$; that is,

$$\frac{a^2}{2 - a^2} + \frac{b^2}{2 - b^2} + \frac{c^2}{2 - c^2} \geq 3 \cdot \frac{\frac{1}{3}}{2 - \frac{1}{3}} = \frac{3}{5}.$$

Note: equality is only possible if $a^2 = b^2 = c^2 = \frac{1}{3}$.

Let p and q be different prime numbers. Solve the following system of equations in the set of integers:

$$\frac{z+p}{x} + \frac{z-p}{y} = q,$$

$$\frac{z+p}{y} - \frac{z-p}{x} = q.$$

Solution

Note that $xy \neq 0$. Clearing the denominators, we get

$$(z+p)y + (z-p)x = qxy,$$
$$(z+p)x - (z-p)y = qxy.$$

By first subtracting the equations, then adding them, we get

$$zy = px, \tag{1}$$
$$xz + py = qxy. \tag{2}$$

Multiplying (2) by z and using (1), we get $xz^2 + xp^2 = qpx^2$, which is equivalent to $z^2 + p^2 = qpx$. This equation implies that p divides z (since p is prime). Thus, p^2 divides qpx. Since $q \neq p$ and q is prime, it then follows that p divides x. Let $x = pa$ and $z = pc$, where a and c are integers and $a \neq 0$. The system is now

$$cy = pa, \tag{3}$$
$$pac + y = qay. \tag{4}$$

From (4), we deduce that a divides y, say $y = ab$. Thus,

$$cb = p, \tag{5}$$
$$pc + b = qab. \tag{6}$$

Since p is prime, it follows from (5) that $(b, c) \in \{\pm(p, 1), \pm(1, p)\}$.

Case 1. $(b, c) = (p, 1)$.
    From (6), we have $qa = 2$, which implies that $q = 2$ and $a = 1$. Then $(x, y, z) = (p, p, p)$, which is a solution for p any odd prime.

Case 2. $(b, c) = (-p, -1)$.
    As in Case 1, we obtain $q = 2$ and $a = 1$. Thus, $(x, y, z) = (p, -p, -p)$.

Case 3. $(b, c) = (1, p)$.
    Equation (6) simplifies to $p^2 + 1 = qa$. Then q must be a prime divisor of $p^2 + 1$, in which case

$$(x, y, z) = \left(\frac{p(p^2+1)}{q}, \frac{p^2+1}{q}, p^2 \right).$$

Case 4. $(b, c) = (-1, -p)$.

As in Case 3, we find that q must divide $p^2 + 1$. Then

$$(x, y, z) \;=\; \left(\frac{p(p^2 + 1)}{q}, \; -\frac{p^2 + 1}{q}, \; -p^2 \right).$$

Therefore, the solutions are $(x, y, z) \in \{(p, p, p), \; (p, -p, -p)\}$, if $q = 2$, for any prime $p > 2$, and

$$(x, y, z) \;\in\; \left\{ \left(\frac{p(p^2 + 1)}{q}, \; \frac{p^2 + 1}{q}, \; p^2 \right), \; \left(\frac{p(p^2 + 1)}{q}, \; -\frac{p^2 + 1}{q}, \; -p^2 \right) \right\},$$

for any primes p and q such that q divides $p^2 + 1$ (which implies that $p \neq q$).

Find all positive integers n such that the equation $x^3 + y^3 + z^3 = nx^2y^2z^2$ has positive integer solutions. Be sure to give a proof.

Solution

Let $n \geq 1$ be an integer such that there exist positive integers x, y, and z satisfying

$$x^3 + y^3 + z^3 = nx^2y^2z^2 . \tag{1}$$

Without loss of generality, we may assume that $x \geq y \geq z$. Then we have $nx^2y^2z^2 \leq 3x^3$, which leads to

$$ny^2z^2 \leq 3x . \tag{2}$$

From (1), we also have $x^2(ny^2z^2 - x) = y^3 + z^3 > 0$, implying that

$$ny^2z^2 \geq x + 1 . \tag{3}$$

Note that $(y^3 - 1)(z^3 - 1) \geq 0$, which leads to $1 + y^3z^3 \geq y^3 + z^3$. Using (1) and then (3), we get

$$1 + y^3z^3 \geq x^2(ny^2z^2 - x) \geq x^2 . \tag{4}$$

Thus, using (2), we deduce that

$$9(1 + y^3z^3) \geq n^2y^4z^4 . \tag{5}$$

Case 1. $y = z = 1$.
From (4), we see that $x = 1$, and the given equation yields $n = 3$.

Case 2. $yz > 1$.
Then $y^4z^4 \geq 2y^3z^3 > 1 + y^3z^3$. This together with (5) implies that $n^2 < 9$. Hence, $n = 1$ or $n = 2$.
For $n = 1$, it is easy to verify that $(x, y, z) = (3, 2, 1)$ is a solution.
For $n = 2$, using (5), we have $y^3z^3(4yz - 9) \leq 9$, which forces $yz \leq 2$. Then $yz = 2$, which implies that $y = 2$ and $z = 1$. It follows that $x \geq 2$, and the given equation reduces to $x^3 + 9 = 8x^2$. Then x^2 must divide 9, which implies that $x = 3$. But it is easy to verify that this is not a solution. Therefore, $n = 2$ is impossible.

Therefore, the desired values for n are $n = 1$ and $n = 3$.

Find all triples of positive integers (p, q, n), with p and q prime, such that

$$p(p + 3) + q(q + 3) = n(n + 3).$$

Solution

First of all, we observe that, for any positive integer m, we have $m(m+3) \equiv 1 \pmod 3$ if $3 \nmid m$, and $m(m+3) \equiv 0 \pmod 3$ if $3 \mid m$. Since we require $p(p+3) + q(q+3) \equiv n(n+3) \pmod 3$, at least one of p and q must be 3. Thus, we may assume, without loss of generality, that $p = 3$. We have $n \geq q + 1$, which means that $n(n+3) - q(q+3) \geq 2q + 4 > 3(3+3)$ unless $q \leq 7$. Checking the primes $q \leq 7$, we find that $q = 2$ and $q = 7$ are the only solutions.

Thus, the only solutions are $(p, q, n) = (2, 3, 4)$, $(p, q, n) = (3, 2, 4)$, $(p, q, n) = (3, 7, 8)$, and $(p, q, n) = (7, 3, 8)$.

Let $0 < a, b, c < 1$. Prove that

$$\frac{a}{1-a} + \frac{b}{1-b} + \frac{c}{1-c} \geq \frac{3\sqrt[3]{abc}}{1 - \sqrt[3]{abc}} .$$

Determine the case of equality.

Solution

The function $f(x) = x/(1-x)$ is convex on the interval $(0,1)$. In what follows, we use Jensen's Inequality and then the AM–GM Inequality:

$$\frac{a}{1-a} + \frac{b}{1-b} + \frac{c}{1-c} \geq 3 \cdot \frac{\frac{a+b+c}{3}}{1 - \frac{a+b+c}{3}} \geq \frac{3\sqrt[3]{abc}}{1 - \sqrt[3]{abc}} .$$

Equality holds if and only if $a = b = c$.

Suppose n is a product of four distinct primes a, b, c, d such that

(a) $a + c = d$;

(b) $a(a + b + c + d) = c(d - b)$;

(c) $1 + bc + d = bd$.

Determine n.

Solution

More generally, we will determine all integer solutions of the system

$$a + c = d, \qquad (1)$$
$$a(a + b + c + d) = c(d - b), \qquad (2)$$
$$1 + bc + d = bd. \qquad (3)$$

If $d = 0$, it is easy to see that the only solutions for (a, b, c, d) are $(1, 1, -1, 0)$ and $(-1, -1, 1, 0)$.

Let us assume $d \neq 0$. From (1) and (2), we have $a(b + 2d) = c(d - b)$, which can be put in the form $b(a + c) = (c - 2a)d$. Using (1) again, we get

$$c = 2a + b. \qquad (4)$$

Then, in view of (1),

$$d = 3a + b. \qquad (5)$$

Using (4) and (5) in (3), we obtain $1 + b(2a + b) + 3a + b = b(3a + b)$, which can be rewritten as $b + 1 = a(b - 3)$. Therefore, $b \neq 3$, and

$$a = \frac{b + 1}{b - 3} = 1 + \frac{4}{b - 3}.$$

Thus, $\frac{4}{b - 3}$ is an integer; that is, $b - 3 \in \{\pm 1, \pm 2, \pm 4\}$. The complete table of solutions can now be easily constructed:

a	b	c	d
1	1	-1	0
-1	-1	1	0
5	4	14	19
-3	2	-4	-7
3	5	11	14
-1	1	-1	-2
2	7	11	13
0	-1	-1	-1

If we now require that a, b, c, d be primes, then $(a, b, c, d) = (2, 7, 11, 13)$, which implies that $n = 2 \cdot 7 \cdot 11 \cdot 13 = 2002$.

Let $a_0, a_1, \ldots$ be an infinite sequence of positive real numbers. Show that $1 + a_n > \sqrt[n]{2}\, a_{n-1}$ for infinitely many positive integers n.

Solution

Suppose, for the purpose of contradiction, that there exists $n_0 \geq 0$ such that $1 + a_n \leq \sqrt[n]{2}\, a_{n-1}$ for all $n > n_0$. We have $\sqrt[n]{2} \leq 1 + \frac{1}{n}$, from Bernoulli's Inequality. Hence, for all $n > n_0$,

$$a_n \leq \frac{n+1}{n}\, a_{n-1} - 1 . \tag{1}$$

We prove by induction on $p \geq 1$ that

$$a_{n_0+p} \leq (n_0 + p + 1)\left(\frac{a_{n_0}}{n_0 + 1} - \sum_{k=n_0+2}^{n_0+p+1} \frac{1}{k} \right). \tag{2}$$

This is true for $p = 1$, because in this case it is just (1) with $n = n_0 + 1$. Now let us assume it holds for some given $p \geq 1$. Using (1) with $n = n_0 + p + 1$ and then applying the induction hypothesis, we get

$$
\begin{aligned}
a_{n_0+p+1} &\leq \frac{n_0 + p + 2}{n_0 + p + 1}\, a_{n_0+p} - 1 \\
&\leq (n_0 + p + 2)\left(\frac{a_{n_0}}{n_0 + 1} - \sum_{k=n_0+2}^{n_0+p+1} \frac{1}{k} \right) - 1 \\
&= (n_0 + p + 2)\left(\frac{a_{n_0}}{n_0 + 1} - \sum_{k=n_0+2}^{n_0+p+2} \frac{1}{k} \right),
\end{aligned}
$$

which ends the induction.

It is well known that $\sum \frac{1}{k}$ diverges to $+\infty$. For sufficiently large p,

$$\sum_{k=n_0+2}^{n_0+p+1} \frac{1}{k} > \frac{a_{n_0}}{n_0 + 1} .$$

For such a p, the inequality (2) forces a_{n_0+p} to be negative, a contradiction.

Let α, β, γ, and τ be positive numbers such that, for all x,

$$\sin \alpha x + \sin \beta x = \sin \gamma x + \sin \tau x .$$

Prove that $\alpha = \gamma$ or $\alpha = \tau$.

Solution

We need only assume $\alpha + \beta \neq 0$.
Differentiating the given identity three times, we obtain

$$\alpha \cos \alpha x + \beta \cos \beta x = \gamma \cos \gamma x + \tau \cos \tau x ,$$
$$\alpha^3 \cos \alpha x + \beta^3 \cos \beta x = \gamma^3 \cos \gamma x + \tau^3 \cos \tau x .$$

In particular, when $x = 0$, we have

$$\alpha + \beta = \gamma + \tau , \tag{1}$$
$$\alpha^3 + \beta^3 = \gamma^3 + \tau^3 . \tag{2}$$

Cubing both sides of (1), we obtain

$$\alpha^3 + \beta^3 + 3\alpha\beta(\alpha + \beta) = \gamma^3 + \tau^3 + 3\gamma\tau(\gamma + \tau) ;$$

hence, $\alpha\beta = \gamma\tau$.
   Consequently,

$$(\alpha - \gamma)(\alpha - \tau) = \alpha^2 - (\gamma + \tau)\alpha + \gamma\tau = \alpha^2 - (\alpha + \beta)\alpha + \alpha\beta = 0 .$$

Therefore, $\alpha = \gamma$ or $\alpha = \tau$.

Find the greatest real number k such that, for any positive a, b, c with $a^2 > bc$,

$$(a^2 - bc)^2 > k(b^2 - ca)(c^2 - ab).$$

Solution

The greatest k is 4.

First suppose that $(a^2 - bc)^2 > k(b^2 - ca)(c^2 - ab)$ whenever $a, b, c > 0$ and $a^2 > bc$. Let $t \in (0, 1)$. Since $1^2 > t \cdot t$, we have

$$(1 - t^2)^2 > k(t^2 - t)(t^2 - t),$$

from which we deduce that

$$\left(\frac{1+t}{t}\right)^2 > k.$$

It follows that

$$k \leq \lim_{t \to 1} \left(\frac{1+t}{t}\right)^2 = 4.$$

Now we will show that $(a^2 - bc)^2 > 4(b^2 - ca)(c^2 - ab)$ whenever $a, b, c > 0$ and $a^2 > bc$. Assume on the contrary that

$$(a^2 - bc)^2 \leq 4(b^2 - ca)(c^2 - ab) \tag{1}$$

for some positive a, b, c such that $a^2 > bc$, and define

$$f(x) = (b^2 - ca)x^2 + (a^2 - bc)x + (c^2 - ab).$$

From (1), either $f(x) \geq 0$ for all real x or $f(x) \leq 0$ for all real x. Actually, the former holds since $f(1) = a^2 + b^2 + c^2 - ab - bc - ca > 0$ (note that $a = b = c$ is excluded by $a^2 > bc$, and so $a^2 + b^2 + c^2 > ab + bc + ca$).

It follows that $b^2 - ca$ is positive. Now, write

$$f(x) = (bx - c)^2 - ag(x) - x(a^2 - bc),$$

where $g(x) = cx^2 - 2ax + b$. Since $a^2 - bc > 0$ and

$$g\left(\frac{c}{b}\right) = \frac{c(c^2 - ab) + b(b^2 - ac)}{b^2} > 0$$

(since $c^2 - ab$ has the same sign as $b^2 - ca$ by (1)), we have $f\left(\frac{c}{b}\right) < 0$, a contradiction. This completes the proof.

The Fibonacci sequence is defined by the following recursion: $f_1 = f_2 = 1$ and $f_n = f_{n-1} + f_{n-2}$ for $n > 2$. Suppose that the positive integers a and b satisfy:

$$\min\left\{\frac{f_n}{f_{n-1}}, \frac{f_{n+1}}{f_n}\right\} \le \frac{a}{b} \le \max\left\{\frac{f_n}{f_{n-1}}, \frac{f_{n+1}}{f_n}\right\}.$$

Prove that $b \ge f_{n+1}$.

Solution

The statement is false, as we can see by choosing $n = 2$ and $a = b = 1$. The correct statement is the one with strict inequalities:

$$\min\left\{\frac{f_n}{f_{n-1}}, \frac{f_{n+1}}{f_n}\right\} < \frac{a}{b} < \max\left\{\frac{f_n}{f_{n-1}}, \frac{f_{n+1}}{f_n}\right\}.$$

Lemma 1. Let x, y, z, t, a, and b be positive integers such that $yz - xt = 1$ and $\frac{x}{y} < \frac{a}{b} < \frac{z}{t}$. Then $b \ge y + t$.

Proof: Since $xb < ay$ and all the numbers are integers, we deduce that $xb \le ay - 1$. Similarly, $ta \le bz - 1$. Therefore,

$$txb \le t(ay - 1) = tay - t \le (bz - 1)y - t = bzy - (y + t),$$

which gives $y + t \le b(yz - tx) = b$.

Lemma 2. For each $n > 1$, we have $f_{n+1}f_{n-1} - f_n^2 = (-1)^n$.

Proof: The proof is by induction on n.

We have $f_3 f_1 - f_2^2 = 2 \cdot 1 - 1^2 = (-1)^2$. Hence, the result is true for $n = 2$.

Assume that the result holds for some given $n > 1$. Then

$$\begin{aligned}
f_{n+2}f_n - f_{n+1}^2 &= (f_{n+1} + f_n)f_n - f_{n+1}(f_n + f_{n-1}) \\
&= -(f_{n+1}f_{n-1} - f_n^2) = -(-1)^n = (-1)^{n+1},
\end{aligned}$$

which completes the induction.

Now assume that

$$\min\left\{\frac{f_n}{f_{n-1}}, \frac{f_{n+1}}{f_n}\right\} < \frac{a}{b} < \max\left\{\frac{f_n}{f_{n-1}}, \frac{f_{n+1}}{f_n}\right\}.$$

Case 1. n is even.

From Lemma 2, we have $f_{n+1}f_{n-1} - f_n^2 = 1 \geq 0$, and therefore, $\frac{f_{n+1}}{f_n} \geq \frac{f_n}{f_{n-1}}$. Thus,

$$\frac{f_n}{f_{n-1}} < \frac{a}{b} < \frac{f_{n+1}}{f_n}.$$

Then, from Lemma 1, we have $b \geq f_n + f_{n-1} = f_{n+1}$, as desired.

Case 2. n is odd.

Arguing as in Case 1, we obtain

$$\frac{f_{n+1}}{f_n} < \frac{a}{b} < \frac{f_n}{f_{n-1}},$$

and the desired conclusion follows once again from Lemma 1.

Let x, y, z be positive real numbers such that $x^2 + y^2 + z^2 = 1$. Prove that

$$x^2 yz + xy^2 z + xyz^2 \leq \tfrac{1}{3}.$$

Solution 1

From the Cauchy–Schwarz Inequality and the hypothesis, we have

$$
\begin{aligned}
x^2 yz + xy^2 z + xyz^2 &= xyz(x + y + z) \\
&\leq xyz(1^2 + 1^2 + 1^2)^{\frac{1}{2}} (x^2 + y^2 + z^2)^{\frac{1}{2}} = \sqrt{3}\, xyz.
\end{aligned}
$$

On the other hand, $\tfrac{1}{3} = \tfrac{1}{3}(x^2 + y^2 + z^2) \geq \sqrt[3]{x^2 y^2 z^2}$, using the AM–GM Inequality; hence, $xyz \leq \frac{1}{3\sqrt{3}}$. The desired inequality follows immediately by combining the two results.

Solution 2

From the identity $(a + b + c)^2 = a^2 + b^2 + c^2 + 2(ab + bc + ca)$ and the well-known inequality $a^2 + b^2 + c^2 \geq ab + bc + ca$, it follows that

$$3(ab + bc + ca) \leq (a + b + c)^2. \tag{1}$$

Setting $a = xy$, $b = yz$, and $c = zx$ in (1) yields

$$3(x^2 yz + xy^2 z + xyz^2) \leq (xy + yz + zx)^2 \leq (x^2 + y^2 + z^2)^2 = 1,$$

from which we get $x^2 yz + xy^2 z + xyz^2 \leq \tfrac{1}{3}$.

Notice that equality holds when $x = y = z = \frac{1}{\sqrt{3}}$.

Find all solutions in positive integers a, b, c to the equation

$$a!b! = a! + b! + c!$$

Solution

If $a \neq b$, then without loss of generality, $a > b$. Then $a!$ does not divide $b!$. Hence, $c! = a!b! - a! - b!$ is not divisible by $a!$. Then $a > c$, and $a!b! = a! + b! + c! < 2a! + b!$, which implies that $(a! - 1)(b! - 2) < 2$. It is easy to check that this gives no solutions.

Therefore, we must have $a = b$. Our equation becomes $a!^2 = 2a! + c!$; hence $a! \mid c!$ and thus $a < c$. Write $c = a + k$ where $k > 0$ is an integer. Division by $a!$ now yields $a! = 2 + (a + 1)(a + 2) \cdots (a + k)$. Then $a! > 2$, which implies that $a > 2$ and $3 \mid a!$. Hence, we must have

$$(a + 1)(a + 2) \cdots (a + k) \equiv 1 \pmod{3} .$$

Then $k < 3$ (otherwise $(a + 1)(a + 2) \cdots (a + k)$ is divisible by 3). If $k = 1$, we get $a! = a + 3$, which yields $a = 3$; then $(a, b, c) = (3, 3, 4)$. If $k = 2$, we get $a! = a^2 + 3a + 4$, which implies that $a \mid 4$; hence, $a = 1$, $a = 2$, or $a = 4$. It is easy to check that these cases fail.

Thus, the only solution is $(a, b, c) = (3, 3, 4)$.

For each integer $n > 1$, let $p(n)$ denote the largest prime factor of n. Determine all triples x, y, z of distinct positive integers satisfying

(i) x, y, z are in arithmetic progression, and

(ii) $p(xyz) \leq 3$.

Solution

Let x, y, and z be distinct positive integers satisfying (i) and (ii). We deduce from (ii) that $x = 2^a \times 3^b$, $y = 2^c \times 3^d$, and $z = 2^e \times 3^f$ where a, b, c, d, e, and f are non-negative integers. Without loss of generality, we may assume that $x < y < z$. Then (i) is equivalent to

$$x + z = 2y. \tag{1}$$

Let $\delta = \gcd(x, y, z)$. Note that $\left(\frac{x}{\delta}, \frac{y}{\delta}, \frac{z}{\delta}\right)$ satisfies (1) and

$$p\left(\frac{x}{\delta} \times \frac{y}{\delta} \times \frac{z}{\delta}\right) \leq p(xyz) \leq 3.$$

Thus, $\left(\frac{x}{\delta}, \frac{y}{\delta}, \frac{z}{\delta}\right)$ is a solution. Moreover, (mx, my, mz) is a solution for each $m = 2^p \times 3^q$, where p and q are non-negative integers. Therefore, we may assume that $\delta = 1$. Then $ace = 0 = bdf$.

From (1), we deduce that x and z have the same parity.

Case 1. x and z are odd.

Then (1) reduces to $3^b + 3^f = 2^{c+1} \times 3^d$, with $f > b$ (since $x < z$). If $b > 0$, then the left side of this equation is divisible by 3, which forces $d > 0$. Then $\delta \geq 3$, which contradicts our assumption that $\delta = 1$. Thus $b = 0$, from which we deduce that $d = 0$. It follows that (1) may be rewritten as $2^{c+1} - 3^f = 1$. But it is well-known (see [1]) that the only integer solution of $2^m - 3^n = 1$ is $(m, n) = (2, 1)$. This leads to $(x, y, z) = (1, 2, 3)$.

Case 2. x and z are even.

Then y must be odd and $d \geq 1$ (otherwise $1 = y > x$). From (1), we deduce that $b = f = 0$. Thus, (1) reduces to $2^a + 2^e = 2 \times 3^d$. Note that $z > y \geq 3$, so that $e \geq 2$. Since the right side of (1) is divisible by 2 but not by 4, this forces $a = 1$. Then (1) is $3^d - 2^{e-1} = 1$. But it is well known (see [1]) that the only integer solutions of $3^m - 2^n = 1$ are $(m, n) = (1, 1)$ or $(m, n) = (2, 3)$. This leads to $(x, y, z) = (2, 3, 4)$ or $(x, y, z) = (2, 9, 16)$.

According to the initial remark, the desired triples are those which have one of the following forms (or their permutations):

$$\left(2^p \times 3^q, 2^{p+1} \times 3^q, 2^p \times 3^{q+1}\right),$$
$$\left(2^{p+1} \times 3^q, 2^p \times 3^{q+2}, 2^{p+2} \times 3^q\right),$$
$$\left(2^{p+1} \times 3^q, 2^p \times 3^{q+2}, 2^{p+4} \times 3^1\right),$$

where p and q are non-negative integers.

Let $f : \mathbb{N} \to \mathbb{N}$ be a permutation of the set $\mathbb{N}$ of all positive integers.

(a) Show that there is an arithmetic progression a, $a + d$, $a + 2d$, where $d > 0$, such that $f(a) < f(a + d) < f(a + 2d)$.

(b) Must there be an arithmetic progression $a, a+d, \ldots, a+2003d$, where $d > 0$, such that $f(a) < f(a + d) < \cdots < f(a + 2003d)$?

[A permutation of $\mathbb{N}$ is a one-to-one function whose image is the whole of $\mathbb{N}$; that is, a function from $\mathbb{N}$ to $\mathbb{N}$ such that for all $m \in \mathbb{N}$ there is a unique $n \in \mathbb{N}$ such that $f(n) = m$.]

Solution

(a) Let $a = f^{-1}(1)$ and $m = f(a+1)$. Then $m \geq 2$. By the pigeonhole principle, the sequence $\{f(a+2^i)\}_{i=0}^{m-1}$ cannot be monotonically decreasing. Hence, there exists k with $0 \leq k \leq m-2$ such that $f(a+2^k) < f(a+2^{k+1})$. Let $d = 2^k$. Then $f(a) < f(a + d) < f(a + 2d)$.

(b) No. In fact, there does not necessarily exist such an arithmetic progression of length 4. Define $f(n) = 4(3^i) - n - 1$ if $3^i \leq n < 3^{i+1}$. Then f is decreasing on the interval $[3^i, 3^{i+1})$ for any integer $i \geq 0$. Suppose that $f(a) < f(a + d) < f(a + 2d) < f(a + 3d)$ for some $a, d \in \mathbb{N}$. Hence, $3^j \leq a < 3^{j+1}$, $3^k \leq a+d < 3^{k+1}$, $3^l \leq a + 2d < 3^{l+1}$, and $3^m \leq a + 3d < 3^{m+1}$, with $0 \leq j < k < l < m$. Therefore,

$$2d = (a + 3d) - (a + d) > 3^m - 3^{k+1} \geq 3^{k+2} - 3^{k+1} = 2(3^{k+1}).$$

Thus, $d > 3^{k+1}$, which contradicts the fact that $d < a + d < 3^{k+1}$.

Find all real k such that the following system of equations has a unique solution:

$$x^2 + y^2 = 2k^2,$$
$$kx - y = 2k.$$

Solution

The system has a unique solution if and only if the line ℓ with equation $kx - y = 2k$ is tangent to the circle γ with equation $x^2 + y^2 = 2k^2$. This condition can be stated equivalently as follows: the radius $\sqrt{2}|k|$ of γ is equal to the distance from ℓ to the origin, which is $\dfrac{|k \cdot 0 - 0 - 2k|}{\sqrt{1+k^2}}$. Thus, we obtain the following condition on k:

$$2k^2 = \frac{4k^2}{1+k^2}.$$

Solving for k, we find that $k \in \{0, 1, -1\}$.

Does there exist a number $q \in \mathbb{N}$ and a prime number $p \in \mathbb{N}$ such that

$$3^p + 7^p = 2 \cdot 5^q \, ?$$

Solution

There is no such pair (p, q).

Assume such numbers p and q exist. Then $p \neq 2$, since $3^2 + 7^2 = 58$ is not of the form $2 \cdot 5^q$. Thus, p is odd. Then

$$2 \cdot 5^q = 3^p + 7^p = (3 + 7) \cdot A = 2 \cdot 5 \cdot A \, ,$$

where

$$A = 3^{p-1} - 3^{p-2} \cdot 7 + 3^{p-3} \cdot 7^2 - \cdots - 3 \cdot 7^{p-2} + 7^{p-1} \, .$$

It follows that $A = 5^{q-1}$ with $q > 1$ (since $A > 1$). Furthermore, since $3 \equiv -2 \pmod{5}$ and $7 \equiv 2 \pmod{5}$, we have

$$A \equiv 2^{p-1} + 2^{p-1} + 2^{p-1} + \cdots + 2^{p-1} \equiv p \cdot 2^{p-1} \pmod{5} \, .$$

As a result, $p \cdot 2^{p-1} \equiv 0 \pmod 5$, and the only possibility is $p = 5$. Since $3^5 + 7^5 = 2 \cdot 5^2 \cdot 341$ is not of the form $2 \cdot 5^q$, the proof is complete.

Determine all pairs (x, y) of real numbers x, y which satisfy

$$x^3 + y^3 \;=\; 7\,,$$
$$xy(x + y) \;=\; -2\,.$$

Solution

Assume that (x, y) is such a pair. Then

$$(x + y)^3 \;=\; x^3 + y^3 + 3xy(x + y) \;=\; 1\,,$$

which leads to $x + y = 1$. Thus, $xy = -2$. It follows that x and y are roots of $X^2 - X - 2 = 0$. Therefore, $(x, y) = (-1, 2)$ or $(2, -1)$, which are solutions of the problem.

Let a, b, and c be positive real numbers such that $a^2 + b^2 + c^2 + abc = 4$. Prove that $a + b + c \leq 3$.

Solution

Without loss of generality, we assume that $0 < a \leq b \leq c$. From $a^2 + b^2 + c^2 + abc = 4$ we deduce that $0 < a \leq 1$, $0 < b < 2$, and $1 \leq c < 2$. Now $c^2 + c(ab) + a^2 + b^2 - 4 = 0$ is a quadratic equation in c and the positive root is

$$c = \tfrac{1}{2}\left(-ab + \sqrt{(4 - a^2)(4 - b^2)}\right).$$

Hence, $a + b + c \leq 3$ if and only if

$$\sqrt{(4 - a^2)(4 - b^2)} \leq 6 - 2a - 2b + ab. \qquad (1)$$

From the AM–GM Inequality, we have

$$\sqrt{(4 - a^2)(4 - b^2)} \leq \tfrac{1}{2}(8 - (a^2 + b^2)).$$

We now prove that

$$\tfrac{1}{2}(8 - (a^2 + b^2)) \leq 6 - 2a - 2b + ab. \qquad (2)$$

This inequality is equivalent to $(a + b)^2 - 4(a + b) + 4 \geq 0$, which factors as $(a + b - 2)^2 \geq 0$. Thus (2) is true. Then (1) is true. Equality is attained when $a = b = c = 1$.

www.ingramcontent.com/pod-product-compliance
Lightning Source LLC
Chambersburg PA
CBHW082133210326

41599CB00031B/5968